THE
HOT
HAND

THE
HOT
HAND

The Mystery and Science of Streaks

BEN COHEN

ch.

CUSTOM
HOUSE

HarperCollins books may be purchased for educational, business, or sales promotional use. For information, please email the Special Markets Department at SPsales@harpercollins.com.

FIRST EDITION

Designed by Leah Carlson-Stanisic

Title page photograph by betibup33/Shutterstock, Inc.

Library of Congress Cataloging-in-Publication Data

Names: Cohen, Ben, 1988- author.
Title: The hot hand : the mystery and science of streaks / Ben Cohen.
Description: First edition. | New York : Custom House, [2020] | Includes
  bibliographical references and index.
Identifiers: LCCN 2019036078 (print) | LCCN 2019036079 (ebook) |
ISBN
  9780062820723 (hardcover) | ISBN 9780062820730 (trade paperback)
| ISBN
  9780062820747 (ebook)
Subjects: LCSH: Chance. | Probabilities. | Cognition. | Fortune.
Classification: LCC BD595 .C645 2020 (print) | LCC BD595 (ebook) |
DDC
  123/.3--dc23
LC record available at https://lccn.loc.gov/2019036078
LC ebook record available at https://lccn.loc.gov/2019036079

ISBN 978-0-06-282072-3

20 21 22 23 24 LSC 10 9 8 7 6 5 4 3 2 1

For my parents, and for Stephanie

CONTENTS

Introduction 1

One   **ON FIRE** 11
"Boomshakalaka."

Two   **THE LAW OF THE HOT HAND** 41
"Unhappy fortune!"

Three **SHUFFLE** 65
"It's just random."

Four   **BET THE FARM** 95
"Principles over patterns."

Five   **WHEEL OF FORTUNE** 125
"It's my life."

Six   **THE FOG** 155
"Be true to the data."

Seven **THE VAN GOGH IN THE ATTIC** 207
"Aha, aha, aha!"

Epilogue 243

An Author's Note on Sources 251

Endnotes. 257

Bibliography 267

Acknowledgments 283

Index. 287

INTRODUCTION

1.

One of the most enchanting moments of my life happened in a meaningless basketball game that nobody else would have any reason to remember. On that day I felt something magical that I have never forgotten. But it would take many years to figure out why. It was all because of a phenomenon that I did not understand, could not have explained, and was not supposed to be real. This book is the story of that seductive idea.

I went to a small high school that barely had enough kids for a varsity basketball team, let alone a junior varsity team. I was on the junior varsity. I started that game on the bench because I started every game on the bench. On this drab winter afternoon, our team walked into a cramped gym, and I went through my normal pregame routine of missing a whole lot more shots than I made. But exactly what happened next I can't really tell you. The sad truth is that I recall almost none of the other material details about this game. The final score, for example. I have no clue. I couldn't tell you which team won, either. The only certainty in my mind is that I must have removed my warm-ups at some point and walked onto the court, because what I'm about to describe couldn't have happened otherwise. It hadn't happened to me before, and it hasn't happened to me since, which is why I still think about it all these years later.

I had the hot hand.

This odd series of events started when I checked into the game after halftime and managed to swish my first shot of the third quarter. I felt good. I swished my next shot. I felt better. I felt like I wanted to shoot again. It was at that point that I swished *another* shot, and it began to dawn on me that I was going to make any shot I dared to take.

As it turned out, contrary to every piece of prior evidence from my pathetic basketball career, the other team was coming to the same bizarre conclusion, and I found myself double-teamed when I touched the ball. I had tricked the other players into believing that I had talent. It was around this time when all the hours I had spent watching basketball on television came in useful. I thought about what someone who was actually good at shooting might do. The next time the ball was in my hands, I pretended I was one of those people. I faked a shot with a surprising amount of confidence for someone who never had a reason to attempt this move before. And it worked! The two defenders flew past me in what seemed like slow motion. Those poor suckers left me enough time to sip espresso before I sent the ball arcing toward the basket and swished one last shot. The whole thing was nothing short of astonishing. In one quarter of one game, I scored more points than I had in my entire life.

I quit playing basketball not long after that game. It was mostly because I was awful. But it was at least partly because I knew that I had already peaked. I would never again experience the rush of having the hot hand.

To have the hot hand is to achieve some elevated state of ability in which you feel briefly superhuman. There is no more pleasurable sensation for humans. Even if you're unfamiliar with basketball, you're probably familiar with this ethereal feeling. The hot hand exists in nearly every industry and touches nearly every person on earth.

What does it mean to have the hot hand in basketball? It's when you've made a few shots in a row, the hoop looks as big as a helipad, and you believe you're more likely to sink your next shot *because* you've made those previous shots. It's those glorious moments when you can't miss that stick with you. But there doesn't have to be a singular definition of the hot hand. You can simply tell when players are hot when you see them ablaze.

I know what that feels like. As much as I watched basketball, I usually found playing basketball to be as enjoyable as driving on the New Jersey Turnpike. But on that night, when I had the hot hand, I was ecstatic. There was no part of me that entertained the silly notion of passing the ball to a teammate. The buzzkill known as "regression to the mean" failed to cross my mind. I was going to shoot almost as soon as I touched the ball, and there was nothing that anybody could have done to stop me. I was hot. For this sublime period of time, I felt like I was defying probability. It would have been totally irrational if not for the fact that everyone on the court believed it was rational. They had seen it before. I had seen it, too. It just had never happened *to* me.

My brush with the hot hand lingered with me long after my basketball career was over. I was still thinking about it when I began doing what so many people do when they can no longer play sports: write about sports. I'm the NBA reporter for the *Wall Street Journal*, and I have written hundreds of stories about basketball, taking advantage of my press credential to gain access to the inner sanctums of the league's thirty teams. It's now my job to watch others catch the hot hand. But one day, while minding my own business and looking for story ideas by reading the latest academic research, I bumped into a ghost from my past. It turned out there were hundreds of scholarly papers about this notion of the hot hand.

I couldn't stop reading these studies. I read them more carefully than I read the lease to my apartment. I found them to be so fascinating because the hot hand was a scientific topic that I didn't have to read a textbook to comprehend. Or at least that's what I thought. I read one paper, then another, and then one more. I was on a hot streak of reading studies about hot streaks. I kept reading and reading until I had plowed through decades of work by economists, psychologists, and statisticians, and there was nothing left for me to read.

Only when I was done reading did I realize why so many people had written so many papers about the hot hand: because there was no such thing as the hot hand.

2.

Pine City, Minnesota, is an obscure speck on the map smack in the middle between Minneapolis and Duluth. I drove to this rural town on a wintry afternoon not too long ago to figure out how the small high school there had become the nation's unlikeliest laboratory of sports innovation. I went to watch the Pine City Dragons hack basketball.

The coach of the Pine City High School basketball team was a history, government, politics, geography, and economics teacher with big eyes and a thick, dark beard named Kyle Allen. When he came to Pine City, the school was known for excellence in the arts, insomuch as it was known for anything at all. Most of his players also played musical instruments. Many of his players were in the choir that sang the national anthem before games. One of his players quit to perform in the winter musical. That player happened to be the team's star player. So there was almost nothing outwardly impressive about the Pine City Dragons.

But they won. They won a lot. They won a lot more than they had any right to win.

It was the way they won, though, that had me curious. I flew to Minnesota, drove to Pine City, and walked back into high school, where I found Allen's players in a classroom inhaling the nutritional benefits of cheese curds and cookies. The same kids busy eating junk food would soon transform into a mighty basketball machine.

The first thing that Kyle Allen did when he moved to Pine City for his first coaching job was blow his entire budget. He was one of thousands of coaches who splurged on technology that created the sort of useful statistics that were previously available only to NBA teams. Pine City's players soon had access to personalized metrics, customized video clips, and more numbers than they saw in their math classes. That profusion of data was the guiding force behind Allen's coaching philosophy. He'd come to Pine City at a peculiar time in the history of sports. The seminal Michael Lewis book *Moneyball* was published when he was still in high school, which put him squarely in the age demographic that straddled a generational divide that was about to roil sports. At issue were the origins of athletic dominance. Kyle Allen understood why he should ground his decisions in data before the data that was relevant to him existed. But once reality caught up, he pounced. He wanted more data. He wanted bigger data. He wanted *better* data.

For his entire life until that point, basketball teams had known precious little about themselves. And not just small high schools in Minnesota. Even teams at the highest levels of the sport were flying blind. There were primitive statistics—basic metrics like points per game—that could give you a sense of value. But there was nothing much deeper than dividing the number of points by the number of games.

That was about to change. Allen spent just about every penny that Pine City's basketball team had as part of his mission to find some value in the numbers. He wanted his team to play smart, and it was quickly becoming clear across basketball that playing smart meant differentiating between the good shots with more value and the bad shots with less value. That simple insight was a breakthrough decades in the making. The good shots were layups and three-pointers. The bad shots were everything in between. The Pine City Dragons became the team that almost never took bad shots.

Their entire strategy was based on maximizing their number of good shots. A typical game for Kyle Allen's merry band of basketball rebels was one in which they took about eighty-five good shots and a few bad shots. In a perfect game, they would take only those shots, and they were closer to perfect than every team in basketball. The bad shots accounted for less than 5 percent of their attempts—lower than any NBA team, any college team, and any known high school team. "In all honesty," Allen said, "that's even higher than we want it to be."

He became a glutton for data after getting his first taste of it. Soon he began the process of quantifying his basketball team. Before long the Pine City Dragons were number one, two, and three in the state record books in one statistical category: most three-pointers in a season. They recorded their stats on iPads, in traditional scorebooks, and on whiteboards in a locker room that carried the unfortunate aroma of teenage boys. They counted the collective number of hours they spent in the gym that summer. They even hired managers to track how much they talked to one another in practice. "Basically," Allen said, as if there were anything basic about it, "everything needs to come down to a number for us."

There was something familiar about the scene inside the Pine City gym on a weeknight that made me unexpectedly nostalgic. I had been in gyms like this one before. I had *played* in gyms like this one before. It was in a gym like this one where I had the hot hand. I should have been able to relate to Kyle Allen's players.

But to watch the Pine City Dragons was to see the future of basketball. They were reinventing the sport into something unrecognizable right in front of me. It felt like being told the sky was green and the grass was blue. I couldn't relate to these kids because I had never thought about which shots had the most value. The only thing I valued was getting home for dinner without embarrassing myself. And it wasn't only because I was a terrible basketball player that I wasn't thinking about this stuff. It was because *nobody* was. And now everybody was.

The kids in Pine City were simply accumulating ideas that had smitten other high school, college, and NBA teams and taking them to a surreal extreme. This counterintuitive strategy to shoot when they were right next to the basket or very far away from the basket but never in between was drilled into their heads until it became intuition. The players no longer needed to be told by their coach to shoot only the good shots. All they had to do was look down at their court. The paint area and land beyond the three-point line were the color of hardwood. The area in between—the section of the court that might as well have been swimming with piranhas—was emerald green. The dreaded part of the floor actually looked different in Pine City. It was yet another reminder of how they wanted to play.

"That's how you *should* play!" one NBA coach said when I told him about this eccentric team I was slightly obsessed with. "Are they better than what they would be?"

They were. The Pine City Dragons had become one of the most fearsome basketball teams in the state of Minnesota. They were harnessing new data, new technology, and new and exciting ways of thinking to reach striking new conclusions about ideas that had long ago been agreed upon. It had been only a decade since I'd mostly humiliated myself in a gym like this one. But in one generation, the game had changed. Everything I thought to be true was very clearly not.

3.

I believed it was serendipity that I had stumbled across the hot hand in my favorite sport. It wasn't. The history of the hot hand has always been rooted in basketball. And so basketball is in this book because it has to be. There is no intellectually honest way to write about the hot hand *without* writing about basketball. The very smart people who have studied the hot hand for a very long time understood that basketball happens to be a wonderful excuse to explore the rest of the world.

But the stories that have always resonated with me are the ones that are not quite about sports, and there are genius scholars and Nobel Prize winners who have devoted their attention to the hot hand in basketball because they weren't just studying basketball. When you start looking for the hot hand, in fact, it becomes hard not to see it everywhere.

That's why I had to make sure I hadn't lost my mind when I read the first scholarly paper about the hot hand that was published in 1985. What made it such a classic work of psychology was its startling conclusion that the hot hand did not exist. This seemed too crazy to be true. As I would soon discover, I was not alone in my shock. The paper was a widely discussed sensation in part because nobody believed it.

We'd all seen the hot hand. We'd all felt the hot hand. The hot hand was burned into our memories. And the appeal of this enticing paper was that it challenged something we all thought to be true. It was a study with a digestible takeaway that forced us to reckon with an eternal question of the human condition: How much should we believe what we see and feel?

The world's brightest academics have been searching for hard evidence of the hot hand ever since. By obsessively looking for proof of something they couldn't find, these people inadvertently turned the hot hand into the Bigfoot of basketball. But those decades of crumpled papers, broken pencils, and deleted spreadsheets only strengthened the case of the original paper. It became clear over time that it was foolish to believe in the hot hand.

Or was it?

That is the mystery at the heart of this book.

We were just beginning to listen to our scientific luminaries and accept that our collective belief in the hot hand may be wrong. And then something incredible happened. It turned out we might have been right to believe in the hot hand after all.

By now you're probably wondering: Is the hot hand real? Yes. But also no. It's complicated. (You might have guessed as much, considering you're about to read an entire book about it.) There are certain situations in which you can take advantage of the hot hand, and there are other scenarios in which allowing the hot hand to guide your behavior can be disastrous. It can be just as costly to indulge the hot hand as it is to ignore the hot hand.

But we'll get there. The story you're about to read is this quest for the hot hand from beginning to end. This is not a book about basketball, but you will have a front-row seat to the most important game of NBA superstar Stephen Curry's career. It's not a book about finance, but you will hear the secrets of a billionaire investor who made his fortune betting against streaks. It's not a book about art or war, but you will meet those who uncovered a long-lost Van Gogh painting and pursued a missing hero of the Holocaust. It's not a book about music, but you will hear from a fabulous composer forgotten by history. It's not a book about literature or medicine, but you will read more than you might have wanted to read about Shakespeare and the plague. It's not a book about technology, but you might think twice before listening to your next Spotify playlist. It's not a book about travel, but you will take a trip to the jungles of the Amazon and to my favorite sugar beet farm on the border of North Dakota and Minnesota.

It's not a book about any of those things. It's a book about all of those things. This is a book about the awesome power of the hot hand. And it begins with man and fire.

One

ON FIRE

"Boomshakalaka."

1.

Mark Turmell was a remarkably odd teenager who became an enormously successful adult for two reasons. The first was that he recognized from a young age what he wanted to do for the rest of his life and never strayed from his ambition. The second explanation for his phenomenal success was that he was a pyromaniac.

When he was a kid in the 1970s, he strolled around his Bay City, Michigan, neighborhood lighting matches along the gutters, walking away, and turning around for a peek, which provided him with the little thrills required to survive childhood. In that fleeting moment when he looked behind him, the unpredictability was so overwhelming that it felt to him like anything was possible. Sometimes there was nothing. Sometimes there was smoldering. And sometimes there was a raging fire. Mark Turmell loved when there was fire.

Turmell managed to keep himself occupied between his infernos by fiddling with computers. He had a friend whose father was a professor at the local community college, and every so often he let the boys play with the computer terminal in his office. Turmell quickly became obsessed with computers. To be more specific, he became obsessed with the video games on computers. He loved them more than he loved fire. Turmell began to sense that he was meant to make video games. He was so confident about his future line of work that he once told his algebra teacher that he didn't need to study for her class because he wouldn't need algebra when he was designing video games. The most amazing part of his pathetic excuse for not doing his homework was that he was actually right.

He was soon going to high school in the morning, taking computer science classes at that local community college in the afternoon, and staying on campus in the computer lab all night. "The only problem with having someone like Mark," one of his professors said, "is that they never want to go home." He was so maniacal about his craft by the time he was fifteen years old that he'd stopped playing his favorite sport, basketball, even though he had the advantage of being one of the taller kids in town, because any time on the basketball court was time that could've been spent playing with computers.

When he decided to buy his own computer, he pooled the money that he'd saved mowing lawns to purchase a brand-new Apple II. Turmell's investment paid off almost immediately. All it took for him to spin a profit was some borderline criminal behavior. Turmell used his Apple II to hack into the community college's network and poke around the sensitive information that schools pay large sums of money to keep private. Once he confessed to his intrusion, the college hired Turmell. It became his job to make sure no one else did what he'd already done. As word of his skills got around town, business got even better for Mark Turmell. Bay City's engineers were so desperate for the expertise of a computer geek that at one point they put software for the local sewage system in the hands of this teenager who didn't have a driver's license yet.

While he was quickly becoming the richest kid in the neighborhood, Turmell wasn't satisfied with his oversight of critical infrastructure. He dreamed of doing bigger things than cleaning the poop of Bay City. He still wanted to make video games, and now there was nothing stopping him. He'd bought the right computer, subscribed to the right magazines, and taught himself the right programming languages. The raw, teenage energy raging in his body kept Turmell awake late at night in his childhood bedroom tinkering on his Apple II. "I kept plugging away waiting for some roadblock that I couldn't surpass," he says. "The roadblock never came."

Turmell's innate talent revealed itself in 1981 with the very first game he made. Anyone who played his shoot-'em-up called *Sneakers* could see it was the work of someone who knew precisely what he was doing, even if what he was supposed to be doing was studying algebra. Turmell shipped a copy of *Sneakers* using a new service called Federal Express to the company that made his favorite games, Sirius Software, not knowing if he would ever hear back. His phone rang a few days later. Sirius wanted to buy his game and guarantee him monthly royalty checks of $10,000. "My dad opened an account, bought some mutual funds, and poured some money in," he says. "I had no idea what was happening." When the most respected Apple II magazine named *Sneakers* one of its most popular releases of the year, the critical and surprising commercial success of Turmell's first game only deepened his resolve. Sirius called again to dangle his dream job: a full-time position making video games. There was no longer any need for school. He moved to California to be with his people.

His reputation on the West Coast preceded him. One of the many people who knew Mark Turmell's name before they met him was the guy who happened to be responsible for his computer. Apple cofounder Steve Wozniak's company had recently gone public and made him a millionaire, and he decided to celebrate by asking his girlfriend to marry him. Wozniak had recently earned his pilot's license and purchased a single-engine plane, and he thought it would be fun to fly down the coast to San Diego to visit her uncle, a jeweler who could help design their rings. But they never made it there. His plane crashed upon takeoff into the parking lot of a nearby skating rink. Wozniak was badly injured and spent the next few months battling a type of amnesia that prevented him from making new memories. He remembered everything that happened before he reached for the throttle and nothing in the five weeks afterward. Only later did he learn that he spent a significant chunk of that time playing Apple II games. When he recovered and felt well enough to finally get married, Wozniak sent a wedding invitation to the man who made one of his favorite Apple II games, a nifty little thing called *Sneakers*. It was the least he could do. Mark Turmell had restored his sanity.

Turmell was something of a celebrity at Wozniak's wedding. At one point he was approached by another young geek.

"Mark Turmell!" he said. "We love your games."

This stranger eventually got around to introducing himself. He had recently founded a software company near Seattle, and he wanted Turmell to come work for him. Would he be interested? "No, man!" Turmell said. He was too busy making video games.

And that was how Mark Turmell blew off Bill Gates. But only in retrospect was turning down the opportunity to be one of the first employees at Microsoft an unfortunate decision. The early 1980s were a great time to be the same brand of geeky as Turmell. He drove a red Porsche convertible. He was profiled in *People* magazine. He received hundreds of fan letters and even some marriage proposals in the mail. Teenage boys wanted to be Mark Turmell. Teenage girls wanted to be with Mark Turmell. By virtue of his talent making video games, he'd turned himself into a bona fide celebrity. He could have worked anywhere he wanted.

The only place anyone who could've worked anywhere he would've wanted to work was Midway. Midway's office was the industry's epicenter of innovation. The companies that worked together in this one

Chicago building were responsible for creating or distributing a staggering number of iconic video games: *Ms. Pac-Man, Mortal Kombat, Galaga,* and so many others that it would be silly to keep listing them because that would mean omitting even more. It's safe to assume that any American arcade game that gobbled your quarters was almost certainly launched by the Midway crew—which soon included the guy who had invented *Sneakers.*

Turmell was so highly valued at Midway that when the company president would walk into his cramped office to ask when his latest game might be done, "I would be able to literally say, 'It'll be done when it's done. And get out of my office,'" Turmell recalls. He could tell his boss to scram because they both understood the harsh reality of their business: Midway sold games to distributors, the distributors sold games to arcades, and the arcades told the distributors how a game was performing. The distributors bought truckloads of that game from Midway if and only if that game was performing well. "There was no amount of marketing, hype, or promotion that could inflect sales," Turmell says. "It was all about cashbox. Nothing else mattered. It had to make cash." And Turmell's gift was for making games that made cash.

The creation of games at Midway followed a meticulous process. Before they were ready to be unleashed on people who would hopefully feed trillions of quarters into the machines, Midway's employees spent hours and hours playing and tweaking these games. Only when Turmell's games had been poked and prodded and probed every which way did they make it to the world outside the Midway office. They didn't travel very far. Their next stop was one of the experimental arcades nearby.

"You don't know what you have in terms of a success or failure until you get it in front of a test audience," Turmell says. "So we'd go sit there and watch."

Mark Turmell was an expert at sitting there and watching. He'd been sitting there and watching for so long that he could predict within a matter of minutes whether one of his games would be a hit. When he visited Midway's favorite test arcade one night in 1992, he realized that his latest creation was going to be the biggest hit of his career. The name of this game was *NBA Jam*.

2.

There is nowhere the world's most talented basketball players would rather come to work than Madison Square Garden. This arena smack in the middle of New York City has been the site of so many divine individual performances over the years that it's amazing it also happens to be the home court of the moribund New York Knicks. But for all those majestic feats in the long history of the game's most hallowed arena, there were three games that stood apart. Three players had returned to the visitors' locker room having scored the most points in the arena's history: Michael Jordan, Kobe Bryant, and LeBron James. The list of people who had conquered the Garden was an exclusive club of all-time NBA greats.

There was nothing to suggest that Wardell Stephen Curry would join their ranks when he walked into the Garden on February 27, 2013, as the evening commuters flooded Penn Station below. But it was strangely appropriate that no one was expecting much from Curry that night. He'd been very good at proving people very wrong for his entire life. Curry went to a small private high school and wasn't supposed to be a big-time college player. When he became a big-time college player, he wasn't supposed to be a good NBA player. When he became a good NBA player, he wasn't supposed to be a great NBA player. He was baby-faced, unassuming, and about as intimidating as a cockapoo.

But there was one thing that Stephen Curry could do better than anybody who had come before him: shoot the basketball. While everyone in the NBA could shoot, no one in the NBA could shoot like him. The most dominant players had always been the ones who made extraordinary things look ordinary. Stephen Curry's genius was making ordinary things look extraordinary.

As he trudged into the Garden that night, Curry was approaching an inflection point in his career. His bum ankles had sidelined him for most of the previous season, and his potential to change the game was still hiding somewhere inside of him. If you think of being a professional basketball player as a normal day job, which in many ways it is and in many more ways it is not, then Curry was similar to most twenty-four-year-olds who'd held the same job at the same company since college. His bosses had given him more responsibilities, and his annual raises and yearly bonuses paid him enough that he didn't bother looking around for better opportunities. In the alternate universe where he sat behind a desk every day, Curry's performance reviews would've been excellent, his recommendations for business school would've been glowing, and his adoring colleagues would've invited him to their weddings. He would've been the ideal corporate employee: highly competent, quietly confident, and extremely useful on the company softball team. He had come closer to working that sort of job than you might think.

When Curry was a college sophomore, his parents bumped into an NBA general manager after one of his games. His mother couldn't help but indulge her curiosity. "Do you think Steph can make it in the NBA?" Sonya Curry asked. There was a reason that not even she could be sure that his future was in basketball. For all the genetic and socioeconomic advantages he'd inherited as the son of an NBA player, there was one severe disadvantage that he couldn't overcome. "On every team he ever played on," says Dell Curry, his father, "he was the smallest guy."

The only way he could hold his own with bigger and better

players, especially as they got even bigger and even better, was by changing the way he shot the basketball. This put Stephen Curry in a deeply ironic predicament. His shot had been his one great skill ever since he'd toyed around with the Fisher-Price baskets in his childhood home. But now someone was telling him that it wasn't good enough. He listened only because that person was his father.

Dell Curry knew that Stephen's strength would soon be his weakness. He could see that his son's low release point meant that anyone taller than him would be able to block his shot. He could also see that everyone was taller than him. Dell took the drastic measure of making Stephen take a break from competitive basketball for a while. In the summer between his sophomore and junior years of high school, when other kids his age were juggling college scholarship offers, Stephen was busy teaching himself to shoot again. By lifting the ball above his head and releasing as he ascended, he was essentially making himself taller. But his learning curve was steep. He took hundreds of shots every day on the court outside his family's stucco two-car garage, where crepe myrtle trees prevented the ball from bouncing into the pool when he missed, and he missed so often that he began to hate shooting. It was a brutal summer that made him miserable. He almost quit basketball altogether.

But that painful summer produced a weapon that Curry would have for the rest of his life. In that summer he became the best shooter the sport of basketball had ever seen. It was that summer that made him a college star and then an NBA player.

He was still in for another rude awakening once he got to the pros. He was good. He wasn't great. When the Warriors played the Knicks in his rookie season, his first time in the Garden as an NBA player, Curry found himself planted firmly on the bench. He could have resigned himself to the fact that he would never be valued properly by the NBA and nobody would have blamed him.

Curry's weapon was the slingshot of basketball. There was an obvious reward for anyone who could wield it: his shots were worth three points instead of two. And not since a biblical shepherd boy named David had the slingshot been used to such a devastating effect. But the slingshot wasn't a bazooka. He still had to be selective about when he shot, where he shot, and why he shot. He couldn't shoot too early in the twenty-four-second shot clock. He couldn't shoot too far behind the three-point line. And he couldn't shoot too much. That restriction was the one constant of his entire career until that February night in Madison Square Garden. Stephen Curry couldn't shoot as much as it made sense for him to shoot.

But what if he could?

3.

There is no getting around it. *NBA Jam* was a spectacularly bizarre rendition of basketball. The characters had cartoonish heads that were bigger than their entire bodies. It was perfectly legal for them to shove, elbow, or pummel the players on the other team. They swished full-court shots and somersaulted above the basket for breathtaking slam dunks. Mark Turmell's creation defied the conventions of sports games because it wasn't supposed to be like existing sports games. His inspiration was a sci-fi game called *Primal Rage*. This video game about basketball was modeled after dinosaurs fighting in a postapocalyptic society.

But from the very beginning, Turmell thought *NBA Jam* had the potential to be a big hit. His careful process for making video games started with turning his colleagues at Midway into guinea pigs, and his test subjects played his games so often they soon needed incentives to keep playing. So they bet. They became compulsive gamblers when they beta tested his games. They usually wagered candy bars. But when it was time for them to troubleshoot *NBA*

Jam, their showdowns were unusually competitive. The developers chose a different form of currency for their bets: cold, hard cash. That was interesting, Turmell thought.

It wasn't long after *NBA Jam* migrated to a local arcade called Dennis' Place for Games that Turmell started hearing that something was wrong with his new game. The *NBA Jam* machine was malfunctioning. Turmell went to the arcade to check for himself. He quickly deduced the problem. It was true that the machine couldn't take any more quarters, but not because the machine was broken. It was because the coin boxes were stuffed. The kids in Dennis' Place for Games were feeding quarters into *NBA Jam* at such a furious pace that employees had to empty the machine every hour so they could keep playing. That was even more interesting, Turmell thought.

The usage statistics as they continued testing the game were off the charts. Every shred of data suggested that *NBA Jam* would be a sensation unlike any video game ever created. But Midway's executives didn't believe the data at first. "We thought the numbers that came back were screwy," said Neil Nicastro, the president of Midway at its peak. "We hadn't yet tested anything that had made that much money."

For a game to be successful in the summer of 1993, it had to earn about $600 per week in the test arcade. There was a thin line between groundbreaking hit and epic flop. If a game made $150, it was a bust. If a game made $1,500, it was a smash. *NBA Jam* made $2,468 in a week when no other game at Dennis' earned more than $750. That number was so ludicrous that Turmell saved a physical copy of the earning report as proof. "Do the math," he says. "It takes ten minutes to play a game. The arcade's open for twelve hours. For that kind of revenue, you have to be playing almost nonstop every day."

The commotion inside Dennis' Place for Games was a preview of the delirium that would infect arcades across the country. Midway needed to sell about two thousand machines to make the game financially worthwhile, and *NBA Jam* would have blown away expectations with ten thousand sales. *NBA Jam* wound up selling more than twenty thousand machines. The mania surrounding Turmell's game was neatly encapsulated by a nasty letter that one out-of-stock distributor wrote to Midway. "Your programmers have created a monster," he wrote.

NBA Jam was *too* successful. It turned out to be one of the most lucrative video games ever made. In less than a year, *NBA Jam* earned $1 billion in quarters.

But why?

There was nothing obvious about *NBA Jam*'s success. The suits who had been skeptical of the numbers never could have imagined that even NBA players would play *NBA Jam*. It wasn't because of the abnormal body types or the acrobatic dunks or even because it felt rebellious and a little bit cool to exhibit such a blatant disregard for the rules of basketball. They became obsessed with *NBA Jam* because of a subtle quirk in the game mechanics.

It was critical to Mark Turmell that each of his games included a goal other than beating the computer. There had to be an elevated state of ability that would compel people to keep stuffing coins into the cashbox. But the inherent problem with sports games was that they were difficult to gamify. They were already games. It was satisfying to win a basketball game, but so what? It wasn't a superpower. Turmell was noodling on this problem one day when he went to Burger King for lunch. He ordered a chicken sandwich with cheese and only cheese. Turmell was always working, even when he was at lunch, and he mentioned his dilemma to another Midway developer named Jamie Rivett. "We need some kind of mode," Turmell said.

By the time Turmell's chicken sandwich with cheese was ready, Rivett had suggested an idea they both knew immediately was brilliant: on-fire mode. They sketched out the details over lunch, walked back to the Midway office, and implemented on-fire mode that afternoon. If a player made two shots in a row, they decided, then he would be heating up. If he made three shots in a row, then he would almost certainly make his next shot. It didn't matter what kind of shot it was. The ball would burst into flames. He would be *on fire*!

This is why Mark Turmell's arcade game was so addicting. Our minds are programmed to search for patterns. He simply programmed a tendency of the human brain that already existed into *NBA Jam*. We see one, two, three shots in a row and intuitively seek out the fourth. We crave order in chaos. Turmell made sure there was a reward for that behavior. He turned the hot hand into *NBA Jam*'s superpower.

Not long after that working lunch at Burger King, Turmell offered the voice-over role for his game to a local comedian named Tim Kitzrow. The script for the gig was two pages. It was exactly the sort of job that no one should have known about. But the test audiences in Dennis' Place for Games became infatuated with Kitzrow's voice. They kept feeding quarters into the *NBA Jam* machines because they wanted to hear a few of the game's catchphrases.

"Boomshakalaka!"

"He's heating up!"

And what they really wanted Kitzrow to bellow were the three words that came next.

"He's on fire!"

Mark Turmell could relate. When he was the age of those kids in the arcade, he had a soft spot for NBA players who caught fire, the ones who made one, two, three shots in a row and everyone in the building knew they were making a fourth. His favorite player was the Detroit Pistons guard Vinnie Johnson, and his nickname was "the Microwave" because he heated up instantly. It wasn't surprising that Turmell idolized Johnson, given his three childhood loves. The first was playing with computers. The second was playing basketball. But it was his third childhood love that explains why he insisted his basketball explode into a fireball when a player was hot. He would have been captivated by *NBA Jam* if he hadn't invented it first.

NBA Jam became unavoidable for boys and girls of a certain age. They played so much that it was as if Mark Turmell had brainwashed a generation of young, impressionable minds into believing the concept of the hot hand. It was systematically drilled into them that anyone who made three shots in a row was almost certainly going to make the fourth. And there was one person who would never be convinced otherwise. This child had an excuse to play *NBA Jam* because his dad was in the game, and he could even pretend to play as himself since they technically shared a name. But nobody called this kid Wardell Curry.

They called him Steph.

4.

The Golden State Warriors were late. There were three buses leaving for Madison Square Garden for their game against the New York Knicks, and Stephen Curry was supposed to be on the second bus. He was always on the second bus. But on this night, for some reason he can't remember, Curry took the third bus. "Which I never do," he tells me a few years later. He regretted the decision almost immediately. The third bus took an illegal turn out of the hotel, and the Warriors were pulled over by unsuspecting traffic cops.

When the bus finally chugged into the bowels of the arena, the players were tired and cranky, and this wasn't entirely the bus driver's fault. The night before had been a rough one for the Warriors. They had lost to the Indiana Pacers in a game that was spoiled by a nasty brawl. They boarded a plane, landed at some ungodly hour, and woke to the news that one of their teammates had been suspended and Curry had been fined for their roles in the fight.

So it had been a lousy day even before Curry found himself stuck on the third bus, dealing with the New York Police Department. But there was nothing he could do about that now. It was time for him to begin his warm-up routine in the Garden. This would be his escape. He started close to the basket, and he kept moving farther and farther back. Finally, as the fans took their seats, he was shooting from several feet behind the three-point line, the strip of paint on every court that was about to redefine the way the game was played.

The three-point line had been introduced to the NBA decades earlier because the biggest people in basketball were too dominant. The sport had become unfair. It discriminated against players like Curry on the basis of their height. With fans tuning out and the game desperate for a jolt, the most democratic solution the NBA could muster was simple math. They slapped a line on the court twenty-three feet, nine inches from the hoop for no reason other than it seemed like the right distance. Any shot within this line would be worth two points. Any shot behind this line would be worth three points.

There is another way of thinking about this radical shift in how basketball was played. The people responsible for the overall health of the NBA were tweaking the algorithm. The word "algorithm" today brings to mind geeks in front of computer screens writing the code that has come to govern our lives. But really an algorithm is a set of rules for solving a problem. When the NBA had a problem, the NBA rewrote the algorithm. The league changed the rules to make the game more exciting and to give players an incentive to stay behind the three-point line—the first in a series of unconnected events that allowed for the hot hand of Stephen Curry.

But the players didn't respond to that incentive right away. In the 1979–1980 season, 3 percent of their shots were three-pointers. Only when their curiosity outweighed their suspicion did NBA players begin to recognize the three-point line as something other than a silly gimmick, and the proportion of three-pointers inched higher until it had reached 22 percent of the total shots in a season by the late 2000s. And then something funny happened. After nearly three decades of steady growth, the percentage of three-point attempts held steady for the next five years. It flatlined. NBA teams were behaving as if they had determined basketball's optimal ratio. The sport had found its equilibrium.

But two things were about to happen that would blow that assumption to bits. The first thing was that Stephen Curry was drafted by a team that he didn't want to draft him: the Golden State Warriors. They were so putrid and their owner was so reviled that he decided to sell the team not long after Curry fell into his lap. The second thing was that a collection of extremely wealthy people with little experience in basketball paid a record fortune to buy the Warriors. They rebuilt their NBA team around the bold notion that they should ignore every orthodoxy of building an NBA team. The construction process took many twists and turns, and there were times when it could have failed, but the eventual dominance of the Warriors can be traced back to one of the most unusual strategies they embraced. It was the notion that the three-point line was a market inefficiency hiding in plain sight.

For almost the entire history of basketball, ever since James Naismith slapped a couple of peach baskets on the wall of a gymnasium and created a sport, the most important area of the court had been around those hoops. The best shots in basketball were always the ones closest to the basket. Or at least that's what people thought. The Warriors weren't sure anymore. "When you can exploit the three-point line," their general manager says, "closer is not necessarily better."

The Warriors came to believe the three-point arc was a boundary in time. Inside the line was the game's past. But the future of basketball was behind the line.

They were one of the first teams to realize they weren't taking nearly enough three-pointers. But the great mystery and baffling paradox of modern basketball is what took so long. At the end of the 2009 season, right before Curry's rookie year, a wonk for ESPN published an article in which he outlined the formula for basketball success: "If you want to exceed expectations, start bombing away from downtown. And if you want to disappoint everyone, stop." He added, "It's no wonder the rate of 3-pointers goes up every season . . . and why it's likely to keep heading in that direction for some time." Except it didn't. At least not for a while. ESPN's basketball expert wasn't such an expert when it came to predicting the behavior of human beings.

How could a group of sophisticated thinkers be so wrong for so long about something that was so important? Pete Carril never understood it. Carril, the legendary coach of Princeton's basketball team, was known as Yoda partly because he looked frighteningly similar to the *Star Wars* character and partly because he was a Jedi master himself. He recognized the value of three-pointers before anyone in his line of work. "I love the three-point shot," Carril once wrote. "You know why? Because it means they're giving us three points for the same shot we used to get two for." It was so obvious that teams should be taking shots that were worth one more point that it was in the name of the shot. That common sense somehow made him a contrarian. But it wasn't *what* he was saying as much as *when* he was saying it. Carril was encouraging his players to take advantage of the three-point line when Curry was still a baby.

The last win of Carril's career came in the first round of the NCAA tournament in 1996. Princeton beat the defending national champion UCLA in an upset that was about as likely as Carril becoming an underwear model. And the world of college basketball reacted appropriately. It went completely bananas. But overlooked in the aftermath was the statistical omen that explained the shocking outcome: more than half of Princeton's shot attempts that night were three-pointers.

As the game was ending, the television cameras panned to the UCLA bench and settled on a player stress-eating his shirt. He was exactly what they were looking for: the face of agony. Many years later, that very same player was hired by an NBA front office, and the team he built would shoot a whole bunch of three-pointers.

His name was Bob Myers. He was the general manager of the Golden State Warriors.

Born and raised in the Bay Area, Myers was a good high school basketball player who never intended to play college basketball. His plan was to join a crew team. The only reason he was on the bench during that game against Princeton was that he'd come to UCLA a few years earlier looking for the rowing coach. It hadn't crossed his mind that he could play basketball there. In fact, before a similar visit to an Ivy League university, he'd written to that school's basketball coach to schedule a meeting. Myers couldn't even get the courtesy of a response. But as he wandered around the UCLA sports complex, he bumped into a basketball coach, who noticed that he was tall and encouraged Myers to attend a tryout. Myers took him up on the invitation. He made the team as a walk-on. He was on the bench as the Bruins won the national championship, and he soon found himself celebrating on the cover of *Sports Illustrated*. That was the thing about Bob Myers. He had a knack for being a part of big things as they happened. "We refer to Bob as our Forrest Gump," his UCLA coach said.

By the time he was a senior, he wasn't just playing for UCLA. He was *starting* for UCLA. The same kid who couldn't get a meeting with an Ivy League university that didn't offer scholarships was now on scholarship as one of the five best players for a basketball powerhouse. And everybody loved Bob. That was actually the headline of a story about him in the school newspaper: "Everybody Loves Bob."

He parlayed that experience, his charming personality, and the handy fact that everybody loved him into a successful career as a sports agent once he graduated. He was good at that, too, and he might have been content negotiating contracts forever. But when his local NBA team was sold, Myers asked for a meeting with the Warriors. He was itching to join another basketball team like the one he'd known in college. As he walked out of his meeting with Joe Lacob, the brash Silicon Valley venture capitalist who'd bought the team, Myers was absolutely positive that he would never be hired by the Warriors. For a while, he was right. Days passed. Weeks passed. Months passed. Myers had the same number of communications with Lacob as he did with that Ivy League coach. And then one day he got an unexpected call.

"Were you serious when you said this might be something you'd be interested in doing?" Lacob said.

Myers quit his job to join the Warriors, and he was quickly promoted to general manager, the top basketball decision-maker for his favorite NBA team. Stephen Curry was one of the foundational pieces of the roster that he inherited. He was the reason Bob Myers would once again be a part of a big thing as it happened.

Myers had always sensed there was a psychological incentive to shoot more three-pointers. He knew from firsthand experience how demoralizing the three-pointer could be for the other team. He was still a little scarred by UCLA losing to Princeton. "I remember viscerally feeling that when you were rooting for a team and the other team hit a three-pointer, it felt like five points," he says. NBA teams had stopped taking more three-pointers by then. That didn't make any sense to the Warriors. It seemed like a good idea to take more of the shots that were worth one more point. "There are analytical reasons to do it," Myers says, "but then I'm not sure many thought it was possible or prudent." But sometimes the most obvious ideas are the most radical. Every now and then they're also the most successful. "What's really interesting in venture capital and doing start-ups is how the whole world can be wrong," Lacob says. "No one really executed a game plan, a team-building architecture, around the three-pointer. Could you win with that?"

It turned out that you could. But first your best shooter had to stop treating his weapon like a slingshot and start using it like a bazooka.

Only because of something beyond his control did all those loosely connected strands of wisdom braid together for Curry in the game against the Knicks. When the NBA reviewed tape of the brawl with the Pacers the night before, Curry had been one of the first players involved in the fight, when he charged a seven-foot-two, 280-pound giant named Roy Hibbert. The outcome was similar to what might happen if a mosquito attempted to tackle a moose. "I didn't even feel him," Hibbert would later say. What saved Curry was his size. He wasn't big enough to do any damage in a fight involving NBA players. For his entire life, Curry had been smaller than everyone on the basketball court, and it had always been a disadvantage. But for this one night it worked to his improbable advantage. The league decided to fine him $35,000 instead of suspending him.

He was amazingly fortunate to lose so much money. The Warriors needed scoring against the Knicks. They had no choice but to free Curry. He was going to shoot more than ever before, and they could only hope that he got hot.

It took him until the second quarter to make his first three-pointer. But one minute later, he made another, longer three-pointer. He was heating up. The next one came a minute later. By any objective measure, it was a bad shot. Curry stole the ball and sprinted across half-court in a straight line on his way to the rim. But instead of continuing toward the basket, which is what almost everyone who had ever played basketball would have done, Curry stopped. He was choosing to stay behind the three-point line. There were two defenders between Curry and the rim who appeared to be shocked by his audacity. Curry was trying a low-percentage shot when a higher-percentage shot was available. He was taking his chances with three instead of accepting two. When the ball dropped through the net, he could hear Tim Kitzrow shouting the *NBA Jam* catchphrase.

Stephen Curry was on fire.

He tried another three-pointer one minute later as he was falling away from the rim several feet behind the arc. This shot was as outrageous as it was ridiculous. It didn't look like it was going in. And then it did. Of course it did! He had the hot hand. "Most locked in I've ever been," he recalls. "Any time I got a glimmer of daylight, I let it go."

When athletes like Curry get hot, they take a puff of the powerful, legal performance-enhancing drug otherwise known as confidence. In the same way that getting a compliment from your boss makes you work harder, hitting one, two, three shots in basketball makes you want to shoot again. The normal chemistry of your brain gets washed away by a flood of dopamine. The frontal lobe begins to act like it's temporarily disconnected from the nervous system. Your muscles melt into Jell-O. You stop thinking. You start behaving intuitively.

But you don't have to be Stephen Curry to be familiar with the feeling of being soaked with adrenaline. One person who could relate to him was Creighton University forward Ethan Wragge. Heavily bearded and slightly overweight, like a lumberjack who got lost on his way to the forest, the closest that Wragge would ever come to an NBA game was buying a ticket. By almost every basketball metric, he was completely mediocre. Except for one. Wragge was a magnificent shooter.

There were practices when he made so many shots in a row that he lost count. It wasn't difficult to figure out how to defend him. The most important part of the scouting report on Wragge—maybe the only important part of the scouting report on Wragge—was knowing not to let him shoot. But there was one night when Creighton played Villanova University and the ball went to Wragge on Creighton's opening possession. He swished it. That one shot was all it took for

Wragge to feel like he was heating up. He was seeing things in slow motion. It was like everyone around him was staggering drunk and he was dead sober. As soon as he hit one shot, he wanted another shot. Wragge tried a deeper shot the next time Creighton had the ball. Again: swish. So he hunted for a third shot. "I feel like it's going in no matter what," he says. He was right. Wragge's third shot went in. So did his fourth shot. And his fifth shot. And his sixth shot. And even his seventh shot. By the time he finally missed, Wragge had scored twenty-one of his team's first twenty-seven points, one of the most amazing shooting exhibitions anyone had ever seen. "It was, like, this automatic, unconscious feeling," he says. "I don't even know how to describe it."

The unscientific name for that automatic, unconscious feeling is "the zone." The zone is a lovely place to be. As it happens, if you needed to describe the zone in two words, you could do worse than "automatic" and "unconscious." The scientists who have actually bothered studying these "flow states" have begun to recognize that acquiring the hot hand is a result of thinking less, not more. The person who pioneered this line of work is a Hungarian American psychologist named Mihaly Csikszentmihalyi, who has spent almost sixty years thinking about the flow state. That's a lot of thinking about flow. What he learned is that being in a flow state is an immensely pleasurable experience. "The way a long-distance swimmer felt when crossing the English Channel was almost identical to the way a chess player felt during a tournament or a climber progressing up a difficult rock face," he wrote. "*What* they did to experience enjoyment varied enormously—the elderly Koreans liked to meditate, the teenage Japanese liked to swarm around in motorcycle gangs—but they described how it felt when they enjoyed themselves in almost identical terms."

The hot hand made them happy. That's the reason Curry, Wragge, and everyone who has felt the hot hand remembers it so fondly.

That night in the Garden was not Curry's first experience with flammability. The first that others remember was when he was six years old and played for a team that was actually called the Flames. But the first that he remembers was in the eighth grade. The Currys had moved to Canada, and Stephen and his younger brother, Seth, enrolled at Queensway Christian College. "We were a small little Christian school where everyone who tried out made the team," says their coach James Lackey. "It was the two of them and a bunch of guys who'd won three games the year before."

Queensway won every game the year that Stephen Curry arrived. He caught fire so frequently that his coach often shook his head in sheer wonder and thought, *What the heck just happened?* Curry's last explosion came in front of an abnormally large crowd for a basketball game between eighth graders. Queensway was playing a team that bullied Curry as if *NBA Jam* rules were in effect. It worked. Down by six points with one minute remaining, Lackey called timeout. He figured his team had no chance to win. Since this was still nominally a middle school basketball game, he wanted to remind his players to keep their composure after the loss and congratulate their opponents when the game was over. Or at least that's what he was planning to say. Curry stopped him in the middle of his sportsmanship lecture.

"We're not losing," he said. "Give me the ball. I'll make sure we win."

"Okay," Lackey said. "I guess that's the play from now on."

They gave Curry the ball. He proceeded to make four three-pointers in the next thirty seconds. Queensway won. Curry's past is littered with so many of these tales that they begin to seem mythological. Lackey swears this one is true. And there is no reason not to believe him: the guy teaches at a Christian school in Canada.

Why did Curry have the hot hand that day? Why did Curry have the hot hand on *any* day? Was it physical? Was it mental? Was it some bowl of cereal that he scarfed down that morning or a lucky seat on the bus that got pulled over on the way to the Garden? Curry himself doesn't know. He can't predict when he'll be in the zone. But he knows he must do everything in his power to remain in that flow state for as long as possible.

"Once it happens," Stephen Curry says, "you have to embrace it."

Curry missed his next shot against the Knicks. If he were in *NBA Jam*, he would have returned to his normal ability. In this real NBA game, Curry didn't take another three-pointer for the rest of the first half. But it wasn't because he suddenly decided that he was lukewarm. It was because the Knicks understood as well as the Warriors that Curry was *still* hot. His teammates refused to touch his right hand because they didn't want to cool him off. "There was nothing anybody could do," said Carmelo Anthony of the Knicks, "except hope he misses."

But there *was* something they could do: not let him shoot. The Knicks sent double-teams at Curry. They trapped him whenever he touched the ball. Their goal was no longer to beat the Warriors. The only thing they cared about was not letting Curry shoot.

Curry knew what it was like to be the sole focus of five other basketball players. When he was a budding star at Davidson College, there was one game when Loyola University Maryland's coach tried to beat Davidson by not letting Curry shoot. His strategy was to double-team Curry no matter where he was on the court and no matter whether he actually had the ball. This coach would rather his team play three-on-four than five-on-five. Curry realized the folly of the plan, stood by himself in the corner, and dragged two Loyola defenders with him. That meant one of his teammates would always be open. Curry could have been eating nachos with fans in the front row and he still would've been helping Davidson. When he noticed that two guys were shadowing him everywhere he went, Curry figured he might as well get to know his babysitters.

"Are you guys really double-teaming me the whole game?" he asked them. They didn't know what to say, so they didn't say anything. The gimmick would have been interesting if it weren't such a complete disaster. Curry was college basketball's leading scorer, and he finished the game against Loyola with zero points. Davidson won in a blowout.

But now the Knicks were more or less going with the Loyola game plan. Curry knew that meant one of his teammates had to be open, and he passed to those open teammates for easy shots. His shooting demanded so much attention that it had become easier for everyone around him to succeed. There's actually a delicious basketball term for this: "gravity." Curry always had the gravity to suck a defense close to him. But his gravity when he had the hot hand made Curry more like a black hole. His momentum warped the game around him. Both teams behaved as if Curry was probably going to make his next three-pointer, and their collective belief in the hot hand was as powerful as the hot hand itself. There was no one on the court who didn't believe in the hot hand. In fact there may not have been anyone in the NBA who didn't believe in the hot hand. "I haven't met that person yet," Curry says.

It wasn't any easier for him to find open shots when the second half started. The Knicks chased him around the court like they were trying to drench him with a bucket of ice water. But his first shot after halftime was all it took to convince him to keep shooting. As soon as he touched the ball, he reminded himself to remain under control when his defender charged at him. He pump-faked—not unlike my own pump fake when I had the hot hand—and watched that defender fly past him. He centered himself, launched the shot, and took in the supreme beauty of his swish.

Curry hadn't cooled off. Once he confirmed that he was scorching, he launched three-pointers that would have earned anyone else a permanent spot on the bench. One from three feet behind the line while double-teamed. One from five feet behind the line. One with a gigantic seven-footer in his face as Curry fell on his butt.

There was a certain sound that accompanied these shots. It began as he released the ball and fans drew a collective breath of anticipation. The pitch rose as their lungs filled. It peaked in a hysterical crescendo as the ball traced a parabola toward the rim. But his shots originated from so far away and followed such a high arc that all these fans ran out of oxygen. That was the noise: the Curry note. It was more recognizable as a shriek.

The last Curry note of the night came at the end of the fourth quarter, when he grabbed a rebound and seemed to be running to the other basket before he even had the ball. He took two dribbles to cross the half-court line. He took one more dribble to slow his momentum. And then he shot. In the millisecond it took for him to levitate, the equation of the possession had tilted Curry's way. His defenders were caught flat-footed. Curry was rising above them. The ball hadn't even swished through the net before he was backpedaling across the court in celebration. He galloped the length of the floor until he was underneath his own basket again. It was as if Curry were literally on fire and needed to extinguish himself. He was *that* hot. "I've never been to quite that place before," he said afterward. "Not ever."

The stunned Knicks fans gave this player from the other team a standing ovation. They didn't know what else to do. Curry had scored fifty-four points—the most points he'd ever scored in a basketball game. In the history of the NBA, no one had taken so many three-pointers and made as many of them. He'd discovered the sweet spot of volume and efficiency.

The three-pointer was no longer a slingshot. Stephen Curry had made it his bazooka.

What happened in the Garden that night wasn't an anomaly. It was an epiphany. His performance emboldened Curry to believe that he could shoot more and that he *should* shoot more. He'd been fully unleashed for the first time, and the results had been astonishing. He'd broken the game.

Curry had the full encouragement of the Warriors' brass to keep shooting after that night. Their decision was part strategy, part stumbling upon something that worked, and part being smart enough to see that Curry would be at his most effective only if he was permitted to do things that nobody had ever done. In his career before that game, he averaged eighteen points, attempting five three-pointers per game. In his career after that game, Curry averaged *twenty-six* points, attempting *ten* three-pointers per game. Curry began shooting as many three-pointers as possible, which was more three-pointers than anyone ever thought possible. There was nowhere on the court that other teams could afford to leave him open. He was a better shooter from thirty to forty feet than the average NBA player was from three to four feet. He turned heaves from near the half-court logo into better shots than slam dunks. He set a record for the most three-pointers in a season, and then he shattered his own record by more than 40 percent. It looked more like a statistical error than a statistical outlier. What he did was almost beyond comprehension. It was the equivalent of Roger Bannister running his four-minute mile in two and a half minutes.

That night in Madison Square Garden when he had the hot hand turned out to be the night that changed Stephen Curry's life. Within two years he was the most valuable player of the NBA. Within three years he was the first unanimous most valuable player in the history of the league. Within four years he was the most influential basketball player alive. The Warriors became an NBA dynasty built around Curry's ability to shoot a basketball. At the peak of his popularity, fans were coming to Warriors games hours early to watch his warmups. But what they really hoped when they paid to see Curry in person was that it would be a night when he got hot. There was simply nothing in sports more thrilling than watching Stephen Curry get hot.

If you ask him for his career breakthroughs, those transcendent moments when he began to feel that he'd achieved what he could only imagine when he was a child playing *NBA Jam*, Curry will tell you about three. There was the time he won his first championship. There was the time he was invited by the White House to golf with Barack Obama. But none of this would have been possible if not for the third moment: the time that he was on fire.

5.

The surfers were catching the last waves before sunset as I walked into a beachfront restaurant on a typically perfect evening near San Diego. The sky was pink. The windows sucked in a soft breeze. The air smelled of salt and grass and sweat. And yet I had the nagging feeling that something was off. I finally realized why. It was such a pleasant summer night that no one was playing with their phones. And the only person who had any reason to be upset about that was the person I was meeting for dinner.

Mark Turmell was now in his midfifties. He was still tall enough to be a basketball player, but he was softer in the belly, and his sandy hair was short and spiky, as if it were apologizing for all those years that his perm fell below his shoulders. Turmell sat down and ordered a cheeseburger with only cheese, just the way he liked his Burger King chicken sandwich. He whipped out his iPhone and scrolled through photos of his wife, whom he had met online, which seemed appropriate. Of course he married someone who was in his life because of a computer. Before long he was scrolling through his apps to show me what he was doing at work these days. It was the same thing he'd been doing for almost forty years. Turmell was still making video games.

He was working for Zynga, the company to blame for addictive games like *Words with Friends* and *FarmVille,* and it was his job to keep people glued to their computers and phones in a way that felt so natural they didn't even notice. He was outstanding at his job. Turmell could've built an actual farm with the productive hours that people had wasted playing Zynga games.

When he was hired by the company, his bosses had begged him to make a "Ville" game. But once again Turmell had another, more ambitious idea.

The first game he released was called *Bubble Safari.* It had all the makings of an arcade classic and would've fit right into Dennis' Place for Games. It was also stupendously dumb. The main character was a monkey named Bubbles, and he was on a mission to rescue his girlfriend, who had been captured by poachers. The only way that Bubbles could sustain himself on his chivalrous expedition was by gathering fruit, and the only way he could gather fruit was by matching pieces that popped the protective bubbles around the fruit. And that was basically it. That was the entire game.

Bubble Safari went live in May 2012. It was the fastest-growing game on Facebook by June. It was more popular than *FarmVille* and *Words with Friends* by July. It spawned *Bubble Safari Ocean*—which was like the original but set in an ocean instead of a jungle and with baby crabs instead of monkeys—and by January that game had become equally addictive. There was a time when more than thirty million people were playing *Bubble Safari.*

That the most popular game on Facebook was about a monkey gathering coconuts and strawberries on his way to rescue another monkey wasn't as fanciful as it sounded—at least not to Turmell. It reminded him of an experience from earlier in his life. *Bubble Safari* had a surprising number of things in common with *NBA Jam.*

"The mechanics are the same," Turmell says. "The key to being successful in this type of market that's so saturated is to have innovation, surprise, and delight around every corner."

NBA Jam had secret characters and crazy dunks. *Bubble Safari* had sticky bombs, paint splats, and double rainbows. And there was one more thing they had in common.

When a *Bubble Safari* player made three matches in a row, Bubbles's ammunition turned the color of a basketball. He was no longer shooting fruit. Now he was spewing flames. Boomshakalaka! Bubbles the monkey was on fire. Turmell swore to himself after the success of *NBA Jam* that he would use the hot hand in every game he developed for the rest of his life. This childhood pyromaniac was still playing with fire.

Was there more at stake for the Golden State Warriors than the kids inside Dennis' Place for Games? Of course there was. But the great insight of Mark Turmell was that Stephen Curry and some pimply teenager with a few quarters in his pocket were really chasing the same thing. They both wanted to take advantage of the rules that controlled their environments to transcend their places in the world. The reward for the *NBA Jam* player was a brief feeling of invincibility and the sound of Tim Kitzrow saying a bunch of funny words. The reward for Curry was an NBA championship.

There was a whole universe of people who had devoted their careers to understanding why NBA players and *NBA Jam* players behaved in similar ways. Mark Turmell had been too busy making video games to know this. He actually didn't know much of anything about this idea called the hot hand. And he didn't know exactly how much he didn't know.

TWO

THE LAW OF THE HOT HAND

"Unhappy fortune!"

1.

It was January 1605, and the queen of England was looking for a good time. When she decided to entertain a foreign visitor with a night at the theater, it seemed like a foolproof plan. The queen was a devoted patron of the arts with a keen appreciation for the playwrights of her day. She was also the queen. One of the perks of being royalty was a seat in the front row of any theater on any night—except for this night.

On this particular evening, there was nothing for her to see. The queen had already seen everything.

There was a reason that even Her Royal Highness was stuck with revivals at this particular moment in British history. It was because the most dependable playwright of her era hadn't written many plays lately. William Shakespeare was in a rut.

But not for long. His fallow period was about to make way for the single most incredible run of Shakespeare's life. Within the span of one year, he wrote *King Lear, Macbeth,* and *Antony and Cleopatra.* Some literary critics believe three of his most enduring plays were released over the course of two short months. Two months! There are juice cleanses that last more than two months. Even if it took him slightly longer, there is no doubt that the period from the beginning of 1605 to the end of 1606 was "a concentrated efflorescence of creative power as strong or stronger than any other in Shakespeare's career," as the scholar J. Leeds Barroll wrote.

Or, as Mark Turmell might say, Shakespeare was on fire.

The same queen who couldn't bear another revival at the beginning of 1605 suddenly had the pleasure of sitting in the front row for not one, not two, but *three* of Shakespeare's greatest plays by the end of 1606. It was such a resplendent stretch of unexpected literary success that it was only natural to wonder exactly what changed.

Was it him? Or was it the world around him?

2.

This is a chapter about how you get hot. It happens to different people in different professions for entirely different reasons. But the process of turning a blip of success into a sustained period of success depends on those same questions about Shakespeare. Was it him? Or was it the world around him? It can be one or the other. But ideally it's both. And that's because the hot hand is not a random occurrence. It's the collision of talent, circumstance, and even a little bit of luck.

Rebecca Clarke certainly had the talent. What she couldn't have known—and what she wouldn't know until it was too late—was if she would get lucky with her circumstance.

Born in the late 1800s in the suburbs of London, Clarke was a viola player who pursued a career as a composer, a radical decision for a woman at the turn of the century. But she was so obviously precocious that one legendary professor went out of his way to cultivate her talent even though he'd never mentored a female composer. She moved to the United States in 1916, and it wasn't long before Clarke had the first hit of her fledgling career. On the afternoon of February 13, 1918, she held a recital in New York's Aeolian Hall. Clarke played three works, including the premiere of a piece for viola and piano called *Morpheus* by a British composer named Anthony Trent, and two duets for viola and cello that she had written. One critic singled out her chops on viola before raving that "as a composer, the young woman likewise shone."

Rebecca Clarke's future was clearly bright. But it was even brighter than anyone in Aeolian Hall could have imagined. That was because there was something they didn't know about her: Rebecca Clarke was also Anthony Trent. She hadn't composed two pieces for the recital. She'd composed three. "I thought it's idiotic to have my name down as composer three times on the program," Clarke said. She chose to invent a pseudonym rather than accept the credit that she deserved.

There was one thing working to Anthony Trent's advantage that Rebecca Clarke would never get to experience for herself: he was a he. Clarke was not especially proud of the piece that she wrote under a man's name, which only made the reaction to *Morpheus* more puzzling to her. "It had much more attention paid to it than the pieces I had written," she said, "which was rather a joke." There was even an article in *Vogue*—a women's magazine!—that mentioned him as one composer worthy of more recognition. A photo of Clarke clutching her viola appeared in the same *Vogue* story, which cautioned that "one should not overlook Miss Clarke's own picturesque compositions." But she wasn't photographed because of those picturesque compositions. It was because she had the great honor of performing an original piece by Anthony Trent. Anthony Trent got the benefit of the doubt because he was not a she.

That kind of slap in the face would've made anyone in her situation want to crawl into a sinkhole. But there was also something undeniably encouraging about this development for Clarke—even if she could see the silver lining only by squinting. The rapturous applause for Anthony Trent was really a confirmation of her major talent.

Her next success was the result of a fantastic opportunity that should have changed her life forever. Clarke was friendly with Elizabeth Sprague Coolidge, a generous benefactor of classical music, who held a chamber music festival every year and sponsored a viola sonata composition contest with a grand prize of $1,000. She encouraged Clarke to enter. It would be a blind competition evaluated solely on the merits of the work and not by the sex of the composer. This seemed like it was almost specifically created for someone with Clarke's musical pedigree. Coolidge was familiar with the inherent disadvantages of being a woman in classical music in the early 1900s. She didn't want anyone else to be. Clarke hadn't yet composed a full sonata, but she couldn't pass on this opportunity, especially not after getting a nudge from Coolidge herself.

The sonata that she wrote was one of seventy-three entries from some of the world's most accomplished composers. That list was whittled down to two finalists and presented to the judges for voting. The musicians on the jury were split right down the middle, and they asked Coolidge to cast the deciding vote. When the jurors opened the envelope to reveal the winner, they read a familiar name: the famous composer Ernest Bloch. But just when the identity of the second composer would have been lost to history, the judges requested that Coolidge open the second envelope, too. They were curious about this composer whom they had determined to be Bloch's equal. They wouldn't have been surprised to see a name like Anthony Trent. What they were not expecting was the name Rebecca Clarke.

"You should have *seen* their faces when they saw it was by a woman," Coolidge said.

Clarke soon entered a piano trio into another one of Coolidge's blind competitions. When she once again placed second, Coolidge was so impressed that she became Clarke's patron. When circumstance shined on her, her talent had sparkled.

There is no doubt in retrospect that this should have been the tipping point of her career. But after the Anthony Trent piece, the viola sonata, and the piano trio, Rebecca Clarke would never have another hit. She basically stopped writing music altogether. Instead she disappeared from public view and kept busy knitting and playing bridge. "I put my things away in a drawer and I got rather embarrassed at even talking about them," she said. Clarke would later refer to her towering achievement in Coolidge's anonymous contest as "that one little whiff of success that I've had in my life."

So what went wrong?

That was more or less the question on Robert Sherman's mind when he made the trip to Clarke's apartment on the Upper West Side of New York City in 1976. It had been nearly six decades since her "one little whiff of success." But if there was anyone who should have known who she was, it was an eminent critic like Sherman, who hosted a radio show that required him to play two hours of classical music every morning. He'd never heard of her.

It was only while putting together a program on a British woman pianist that he learned, to his great shock, that another British woman who had worked with this pianist was still alive and living not too far away. By the time he called Clarke, she was eighty-nine years old and used a walker to get around. She shuffled over to her closet and reached for a program commemorating a performance she'd given with this pianist whom Sherman had come to hear about. He couldn't help but notice there were pieces on the program written by the woman who was now sitting across from him. He'd known that Clarke was a violist. He hadn't known that she was a composer, too. "Oh, long ago," Clarke said. "Nobody remembers."

Only when he prodded did she proceed to tell Sherman about her forgotten past. He very quickly realized that there was an even better story staring him straight in the face. He scheduled another meeting with Clarke—and this one would be solely about Clarke.

"Why did you stop writing songs?" he said.

"Well, that's the $64,000 question, isn't it?" she replied.

Clarke had the talent to be one of the great composers of the twentieth century. But her circumstances never coincided with her talent. Clarke was alive at the wrong time. She wasn't in an environment that allowed her to capitalize on the hot hand. Her family members resented her line of work and mocked the very creations that would one day be celebrated. "There was a lot of giggling underneath the surface about her music," said one relative. "In the family, it was considered to be absolutely ridiculous." But really their objections boiled down to the fact that Rebecca Clarke was a woman, and composing beautiful music was thought to be a frivolous activity for women. She was never able to maintain a singular focus on her work, even with the support of a wealthy patron like Coolidge, because of these detrimental conditions that she encountered. Her career stalled at the moment it should have exploded, and she begrudgingly accepted her circumstances.

"I didn't—I seemed to lose my interest in—I—" Clarke attempted to explain to Sherman. "I can't really quite tell you all about it."

But there was once a time when she could. She even described the enigmatic phenomenon of the hot hand in her unpublished memoir. "Every now and then, in the middle of struggling with some problem, everything would fall into place with a suddenness almost like switching on an electric light," she wrote. "At these moments, though I had no illusions whatever about the value of my work, I was flooded with a wonderful feeling of potential power. A miracle made anything seem possible. Every composer, or writer, or painter too for that matter, however obscure, is surely familiar with this sensation. It is a glorious one. I know of almost nothing equal to it."

Clarke was still clinging to that memory decades after she'd felt it for the last time.

"There's nothing in the world more thrilling," she told Sherman. "But you can't do it unless—at least I can't; maybe that's where a woman's different—I can't do it unless it's the first thing I think of every morning when I wake and the last thing I think of every night before I go to sleep."

"You need that intensity and that concentration," he said.

"Yes," she explained. "I can't do it otherwise."

Her feelings of deficiency ran so deep that instead of being revered for her accomplishments she wanted to forget them altogether. What she *hadn't* accomplished hurt too much. "Most people don't even know that I ever did any composing because I didn't like talking about it," she said.

Sherman shared Clarke's story with a violist, pianist, and chamber trio who looked at the scores that she'd fished out of her closet and performed her three known pieces on the radio. When the interview aired on August 30, 1976, it was accompanied by music that almost nobody alive had ever heard. "It was nice, and her friends enjoyed it, but the important thing was that it went *beyond*," Sherman says. "It went beyond in a way that none of us could have ever imagined." The violists and pianists and chamber trios listening at home were astounded. They didn't know who she was. They didn't really care, either. Enough time had passed that they were listening to her music with the same lack of bias as the judges in a blind competition. They soon came to a similar conclusion: Rebecca Clarke was an exemplary composer.

The pieces that had been sitting at the bottom of her closet for decades came back into circulation. Soon there were fresh recordings of her sonata and trio. Less than a year after the radio show, Clarke's viola sonata was performed in Lincoln Center. "Had she not been a woman composer when such phenomena were not taken very seriously, Miss Clarke might be heard more today," wrote a *New York Times* critic in a glowing review that called her sonata "a lovely piece full of arresting melodic ideas that often strike a note of genuine passion and originality." There would eventually be a collection of interviews with her and essays about her called *A Rebecca Clarke Reader*, edited by a musicologist named Liane Curtis, the president of The Rebecca Clarke Society. What had begun as a good deed for a nice old lady had become a full-blown comeback for a deeply misvalued artist. "Before she died," Sherman says, "there was a total revival of Rebecca Clarke's pieces." The most amazing part of all this was that he'd inadvertently sparked the resurgence of a forgotten composer all because he needed tape to fill a radio show that wasn't even supposed to be about her.

"It was a confluence of accidents," he says, "and fortuitous circumstances."

There's that word again: "circumstance." Only right before she died in 1979 did Clarke begin to take her rightful place in the canon. It was through no fault of her own that she wasn't able to take advantage of her hot hand. Who knows what might've happened if conditions had evolved in Clarke's favor instead of conspiring against her? There are composers who spend their whole lives toiling in obscurity and die without anybody knowing their names or hearing their music. Their lack of circumstances is less tragic because most of them don't have the talent. She did. And the inevitability of her fate was the sort of thing that she couldn't help but think about. She even indulged herself by dabbling in tarot cards. She even read her own fortune every now and then.

"I did my own a number of times," Rebecca Clarke said. "They came out different every time."

3.

A century after Rebecca Clarke performed her first concert in the United States, a statistical physicist named Dashun Wang found himself thinking about people like her, even though he had no clue who she was. He was more interested in someone whose name he did know.

Wang focused on the year in which Albert Einstein managed to produce his research on the photoelectric effect that would later win him a Nobel Prize. At that point Einstein could have called it quits and taught himself how to yodel. Instead he published his theory of special relativity, a study of Brownian motion, and the most famous equation in the history of science: $E = mc^2$. He packed a career's worth of intellectual achievements into a few months. What he did in 1905 is now known simply as "the year of miracles."

But as he contemplated this run, Dashun Wang began to wonder how miraculous that year really was. By thinking about Einstein, he made an important scientific discovery of his own.

Let's call it the law of the hot hand.

Wang believed it was not a coincidence that Einstein's best work came in bunches or that Clarke had three straight triumphs. That's how creativity works. Success is streaky. Hits are clustered. And this heightened state of ability is the key to understanding the lasting contributions of everyone from scientists to movie directors. Their careers are defined by their hot-hand periods. "What happens in the hot-hand period," Wang says, "is what we remember."

The first movie that one of those movie directors actually directed was a satirical documentary about a fictional, mediocre British rock band. Rob Reiner dragged his twenty-minute demo cut from studio to studio and kept being told that his mockumentary called *This Is Spinal Tap* would never work. He finally got the minimal amount of funding he needed from one of his father's oldest friends and shot the whole film in five weeks on a shoestring budget. The movie was at best a modest success at the box office, but it was a smash with the critics. Roger Ebert gave it four stars and called *This Is Spinal Tap* "one of the funniest, most intelligent, most original films of the year." That review alone made Rob Reiner a movie director. All he had to do next was direct another movie.

The logical next step for someone in his position would have been to play it safe and ease his way into the mainstream. But he decided to make another movie that he wasn't supposed to make. It was built around a bolder premise: a romantic comedy that dared to treat teenagers like adults. *The Sure Thing* was also a critical hit, and this time his film made some money. He'd earned the runway to make a third movie. And what did Rob Reiner do? He picked *another* movie he wasn't supposed to make. The collective wisdom of the show-business crowd suggested that *Stand by Me* was bound for disaster. It was based on a story by horror novelist Stephen King, but it wasn't a horror movie. It was almost as if Reiner and King were intentionally trying to alienate their most loyal fans. As if that weren't enough to sink the film, Reiner basically cast a bunch of unknowns. And still the movie became a monster at the box office.

Now there was no longer any doubt about his directing ability. Rob Reiner was critically successful and commercially bankable. That was as powerful a Hollywood combination as peanut butter and jelly. This was finally starting to become clear to the movie studios that kept passing on Reiner's projects only to watch the positive reviews and piles of money come pouring in afterward. But who could blame them? His movies were delightful contradictions. "He is successful not because he made movies that everyone expected to be a hit," a newspaper reporter once wrote about Reiner, "but because he made movies that no one expected to be a hit." So you might think that some enterprising producers would have sold their kidneys to put their names on a Reiner movie. But no. One of Reiner's exchanges with a studio executive went like this:

"We love your films," the studio executive said. "What do you want to do next?"

"You don't want to do what I want to do," Reiner replied.

"No, that's not true. I want to do what you want to do."

"No, you want me to do what you want to do."

"No, I want to do what you want to do."

The studio executive finally asked Reiner to name his next film.

"*The Princess Bride.*"

"Well, anything but that."

Let's go back to what happened when Stephen Curry had the hot hand. His team started running plays to get him more shots, and his coaches demanded that he keep shooting. The result of his scoring was that it became more likely that Curry would score. A similar thing happens if you're a director with the hot hand. Screenwriters, actors, and studios want to work with you. They want to assist you. They want to make sure you shoot. You get better opportunities because you're hot. Success begets success. That is the simple power of the hot hand.

Remember what Curry said about the hot hand? Once it happens, you have to embrace it. Rebecca Clarke didn't have the means to embrace it. It would turn out that Rob Reiner did.

Curry was a different player when he was on fire. He took longer shots. He took harder shots. He took shots that he never would have taken if he didn't have a hunch that he was hot. There was even a name for these absurd shots that Curry took knowing that nobody could fault him if he missed. They were called "heat checks." Rob Reiner's heat check was *The Princess Bride.*

The Princess Bride was Hollywood's great white whale. It was a fairy tale with fight scenes and true love, a children's movie for adults that was silly and sweet, a film that managed to fuse romance, suspense, comedy, and drama. It was also a riddle haunted by a curse. Even after his three successes in a row, Reiner knew it would be tricky to make *The Princess Bride*. He didn't know that it had proven impossible until that point, and he might have chosen a different project altogether if he'd been aware of the movie's daunting history.

François Truffaut and Norman Jewison had tried to make *The Princess Bride*. Robert Redford tried to make and star in *The Princess Bride*. The legendary screenwriter William Goldman liked to tell people that multiple studio heads had been fired immediately after promising to make his film. This was odd. Any studio that wanted to be a hit factory would have been wise to follow a simple formula: make Goldman movies. This was the guy who wrote *Butch Cassidy and the Sundance Kid* and *All the President's Men*. His material was so valuable that by the early 1980s no one would have blamed Hollywood executives for rifling through his trash and buying the rights to his grocery lists. But his stature only made the convoluted saga of *The Princess Bride* more baffling.

There is so much that has to go right to make a movie that it's a miracle any movie ever gets made. In theory it should have been easier for Rebecca Clarke to compose an entire symphony than it was for Rob Reiner to direct *The Princess Bride*. After all, if composing music is an individual pursuit, then directing a movie is a collective endeavor. But this is when it came in handy that Reiner was aware that he had the hot hand. He used it to bend circumstance his way.

When he decided to make this movie, to risk his directing career and spend the capital that he'd earned making a few hit movies, Reiner went to convince Goldman that he was worthy. He was petrified to ring the doorbell of his apartment. "*The Princess Bride* is my favorite thing I've ever written," Goldman said after he opened the door. "I want it on my tombstone." Reiner discussed his vision for the film and showed Goldman cuts of his previous movies. He would later call the meeting in which he earned Goldman's blessing the greatest moment of his directing career. Once he had permission, there was one more thing Reiner needed: money. *The Princess Bride*'s screenplay called for swordplay, gigantic rodents, and torture chambers, as well as a cast of characters that included a Sicilian hunchback, the most beautiful woman in the world, and a giant. This was epic stuff. The problem with epic stuff was that it was also expensive. Reiner finally convinced Norman Lear, the *All in the Family* creator who happened to be a friend of Reiner's father, to open his checkbook and lend him enough cash to make the movie. Only then did Twentieth Century Fox come along and agree to distribute Reiner's project—even if that project was *The Princess Bride*. Reiner still laments how difficult it was to make *The Princess Bride*. The hot hand was the only reason that it was even remotely feasible.

We're all lucky for it. If there were some way to quantify the number of people who have seen and enjoyed any given movie, *The Princess Bride* would rank at the top of the list. One of the closest approximations to that metric is a movie's CinemaScore grade. CinemaScore is a research firm with movie data going back several decades, and this combination of longevity and reliability gives its grades a certain weight. If a movie has a good CinemaScore grade, it's probably good. If a movie has a great CinemaScore grade, it's almost certainly destined for greatness. But if a movie gets an A+ grade, it's an instant classic.

It's a profound achievement to make even one movie worthy of the A+. Only ten directors have two A+ movies. But there is one director with *three* A+ movies. They were released within a span of five years and have almost nothing in common except for the man in the director's chair. Rob Reiner made *A Few Good Men* in 1992. He made *When Harry Met Sally . . .* in 1989. And his run of A+ movies started in 1987 with the making of a movie that no one wanted him to make.

That movie? It was *The Princess Bride*.

4.

To understand what Dashun Wang and his collaborators found, it helps to understand how they found it. Wang's team wanted to put some numbers behind fuzzy concepts like artistic taste, scholarly impact, and whether a movie is good, and they went searching for objective data that could help them quantify the subjective. The point of their research was not to compare academics to directors. It was not even to compare academics with academics or directors with directors. It was to compare the works of Einstein with the other works of Einstein. They wanted to pinpoint when creative types peak. The only way they could do that was by comparing their subjects with themselves.

The data they collected was sufficiently large to make some interesting conclusions. They looked at the auction prices for three thousand artists, the Google Scholar and Web of Science citations of twenty thousand academics, and the IMDB movie ratings of six thousand directors. Once they looked at those numbers, they found themselves staring at a surprising pattern: 91 percent of financially successful artists, 90 percent of published scientists, and 82 percent of directors whose films reached theaters had at least one hot-hand period in their careers. They had caught fire. The most expensive paintings, influential research, and beloved films were not independent events. They were the by-products of creative streaks.

While he found evidence for the hot hand only among artists, scholars, and directors, Wang is convinced that he would find it in any other industry. He believes it's universal. When people had the hot hand, their quantity of work might have been the same, but the *quality* of that work was empirically higher. These were prolonged stretches of professional success in which people outperformed even their own expectations. They took advantage of their resources when they were hot to get hotter. These cultural luminaries and scholarly dignitaries were the best versions of themselves when they had the hot hand. This was fundamentally different from the fleeting rush of the hot hand in basketball. It wasn't a matter of short-term momentum. The peaks of their careers lasted anywhere from three to five years, and the way their hits built on one another meant there were long-term effects to getting hot.

"If I know your best work, I know when your second best will be and when your third best will be," Wang says. "That's your hot-hand period." But it's not linear. It's jagged. "You're progressing along with a certain level of performance, and then all of a sudden your performance elevates to another level," Wang explains. "You're not yourself anymore. You're not producing more than you expect. But what you produce in that period is much, much better." Maybe the most riveting thing he found was actually something he didn't find: there is no way yet to predict *when* someone is on the verge of such a streak. "Your hot streak can come at any time," Wang claims. "What I learned from my own research is actually rather uplifting. Because the hot streak can start with any work, the only sure way to prevent it is to stop publishing. If you keep going, your hot streak may be yet to come."

But how do you know if you've already had your hot hand? You don't. You can't! If you were to ask Wang for advice, he might tell you that it doesn't matter what your circumstance is. It doesn't matter if you're a Rebecca Clarke or a Rob Reiner. It doesn't even matter if your hot-hand period is on the horizon or if it has already passed.

"The answer is the same," Wang says. "You should keep going."

Wang is such an unfailing optimist about the hot hand that he makes it easy to forget that behaving in this manner can backfire. It's a supremely risky philosophy. Confidence can become arrogance. Arrogance can become ignorance. Your internal ability can get you only so far. That doesn't mean you should quit your beer-league basketball team just because you won't make it to the NBA. But it does mean that you shouldn't delude yourself. Stephen Curry recognized the conditions were ripe for a hot hand. Rebecca Clarke realized they weren't. That is the loophole in Wang's law of the hot hand. There are many industries where your internal cadence is at the mercy of external forces. Sometimes those external forces break your way and the result is magic like *The Princess Bride*. Sometimes those external forces crush ambition. Talent is important, but circumstance is imperative, and circumstance beats talent when talent doesn't have circumstance.

But the reason to put some faith in Dashun Wang and trust the law of the hot hand is that it's not out of the realm of possibility for circumstance to appear when talent least expects it.

In fact it's happened before.

5.

One summer day in 1564, before Dashun Wang, before Rob Reiner, long before Rebecca Clarke and even before Queen Anne, a small village in the English countryside was rattled by the sudden death of a weaver's apprentice. The local tragedy was immortalized in the margins of the town's records. Next to the name of the weaver's apprentice were three ominous Latin words: *Hic incipit pestis.*

"Here begins the plague."

The plague wiped out a sizable portion of this particular town. It was an indiscriminate killer. The weaver's apprentice was the first of more than two hundred people to die over the next six months.

Who lived and who died was seemingly a matter of chance. The plague could decimate one family and spare the family next door. In one house on Henley Street was a young couple who had already lost two children to previous waves of the plague, and their newborn son was three months old when they locked their doors and sealed their windows to keep the plague from invading their home again. They knew from their unfortunate experience that infants were especially vulnerable to this morbid disease. They understood better than perhaps anyone on Henley Street that it would be a miracle if he survived. Seven of ten babies died in plague years. It was as if every family were flipping a coin unfairly weighted toward heads and betting a child's life on tails.

But when the plague was done with this small village in the English countryside, a little town called Stratford-upon-Avon, the couple breathed a sigh of relief that their young boy was still alive. William Shakespeare was a miracle who grew up to do miraculous things.

There's a possibility that Shakespeare developed some kind of immunity to the plague because of his exposure when he was an infant, but that speculation began only centuries later and only because the plague was a constant nuisance to Shakespeare. "Plague was the single most powerful force shaping his life and those of his contemporaries," wrote Jonathan Bate, one of his biographers.

Shakespeare was around the plague enough to recognize its symptoms. First the body temperature would spike. Next came the headache that would spread to the back, the legs, the groin, the armpits, and the neck. Before long everything would hurt. Anyone who tried to walk at this point would have looked and sounded like he'd chugged a bottle of tequila. His breathing would have been so labored that he wouldn't have been able to talk without slurring his words. It would only get worse from there. The skin would become a patch of carbuncles—even the words associated with the plague were gruesome—and by that point the outcome would be inevitable. The last stage of torture was the brain crying uncle. The victim endured the last few hours of his life in a state of madness. The whole thing was wretched enough to spend your life worrying about how you might die.

The plague was naturally a taboo subject for much of Shakespeare's writing career. Even when it was the only thing on anybody's mind, nobody could bring himself to speak about it. Londoners went to the city's playhouses so they could temporarily escape their dread of the plague. A play *about* the plague had the appeal of watching a movie about a plane crash while thirty-five thousand feet in the air.

But the plague was also Shakespeare's secret weapon. He didn't ignore it. He turned his enviable talent and his lamentable circumstance into the hot hand.

And that brings us to the macabre history of *Romeo and Juliet*.

It's basically impossible to appreciate the truly bonkers nature of this play when you read it for the first time. You probably remember the basics of the plot: Romeo and Juliet are born into rival families; Romeo and Juliet fall in love; Romeo and Juliet die. But do you remember *how* any of that happens? Maybe not. And did you know the plague is what ultimately drives Romeo and Juliet apart? I bet you didn't. Perhaps you vaguely recall the only explicit mention of plague in the entire play: "A plague o' both your houses!" But the plague is actually everywhere in *Romeo and Juliet*.

Let's refresh your memory of Shakespearean dramas. In case you don't remember, there's a death in Act III of *Romeo and Juliet*. A murder! Romeo has killed his rival, Tybalt, who happens to be the cousin of Juliet. At this point she's supposed to marry Paris, but she's actually in love with Romeo, which is a problem because Romeo's family is the sworn enemy of Juliet's family. It's also a problem because Romeo is now banished after killing Tybalt. Juliet doesn't know what to do. She turns to her spiritual authority, Friar Laurence, who has already decided there is one way and only one way to bring the

Montagues and Capulets together. To end their blood feud, he will have to marry Romeo and Juliet. Friar Laurence's new plan requires Juliet to drink a potion that will put her to sleep for so long that her family will have no choice but to conclude that she is dead. At the same time, Friar Laurence writes a letter to Romeo explaining the harebrained scheme, and Friar John will deliver it to the town of Mantua. The letter instructs Romeo to sneak back to her open coffin and steal Juliet so they can live happily ever after.

It was a pretty terrible plan that worked out pretty terribly—but not for the reasons you might expect. Juliet drinks the potion. Her family concludes she's dead. Romeo sneaks back to see her. So far, so good. But the whole thing unravels because of what should have been the most reliable part of this ridiculous plan: Friar John never makes it to Mantua, and Friar Laurence's letter never makes it to Romeo.

What happens next is a series of highly unfortunate events. Romeo thinks Juliet is dead. He kills himself. Juliet wakes up from her fake death and learns that Romeo is dead for real. She kills herself. *For never was a story of more woe / Than this of Juliet and her Romeo.*

But let's rewind a few scenes. Let's read how Friar John explains to Friar Laurence why he never reached Mantua. Let's figure out how the whole foolish scheme fell apart.

FRIAR LAURENCE
Welcome from Mantua. What says Romeo?
Or, if his mind be writ, give me his letter.

FRIAR JOHN
Going to find a barefoot brother out,
One of our order, to associate me,
Here in this city visiting the sick,
And finding him, the searchers of the town,

Suspecting that we both were in a house
Where the infectious pestilence did reign,
Sealed up the doors and would not let us forth.
So that my speed to Mantua there was stayed.

FRIAR LAURENCE
Who bare my letter, then, to Romeo?

FRIAR JOHN
I could not send it—here it is again—
Nor get a messenger to bring it thee,
So fearful were they of infection.

FRIAR LAURENCE
Unhappy fortune!

Look again: *Where the infectious pestilence did reign, / Sealed up the doors . . . / So fearful were they of infection.*

Why didn't Friar John deliver Friar Laurence's letter to Romeo? Because of the plague. The plague is the plot twist that turns the most famous love story ever told into a tragedy.

Friar John never makes it to Mantua with the letter for Romeo because he gets stuck in quarantine, and no one in Shakespeare's time would have dared to question such a restriction. They knew that defying quarantine made you eligible for a whipping and walking around town with plague sores could be punished by execution. It was a regrettable way to die: you survived the plague only to be killed for it. This is why Friar John doesn't understand how Friar Laurence could be so upset with him. Of course he doesn't. He doesn't realize it's because the plague is about to kill Romeo and Juliet.

The exchange between the friars amounts to twenty-four lines. It's one of the shortest scenes in the entire play. It's over before most kids reading it in high school realize what has happened. And yet it's essential that we do. The whole play turns on this one scene. You might be wondering how the plague could be pulling the strings of a Shakespeare play and you might not have known until this very moment. As it turns out, that was the point. Shakespeare was being purposefully obtuse. He wrote in veiled language because the subtext would have been obvious back then. He didn't have to belabor the point. The plague was the Shakespearean equivalent of ending a tweet with "Sad!" There was no need for any sort of further explanation. "It was omnipresent," says Columbia University professor James Shapiro. "Everybody at the time would have known exactly what those one or two lines meant."

Romeo and Juliet would not be the last time that Shakespeare used the plague to his advantage. The rule of thumb until not too long ago was that Shakespeare wrote two plays every year. But when Shapiro began his life's study of the playwright, he deduced that his fellow literary scholars were not exactly statisticians. They had come to that number simply by dividing the number of plays he wrote by the number of years in which he wrote them. According to their calculations, if Shakespeare wrote ten plays in five years, he wrote two plays a year. The actual chronology of those plays had been mostly ignored ever since Shakespeare's contemporaries organized his First Folio not by year but by category: comedy, tragedy, or history. This dubious math went unchallenged for hundreds of years. By the time Shapiro became a professor, the notion that Shakespeare wrote two plays every year was close to gospel. But there was a rub. And the rub is that it wasn't remotely true.

"It turns out Shakespeare always tended to write in inspired bunches," Shapiro says. "It's something that took me a while to wrap my head around simply because I always kind of believed the unsubstantiated claims that he was churning out two plays a year. But that's never what he did."

Shakespeare ran hot and cold. His plays were not spread over the course of his career. They were clustered. If he'd studied playwrights instead of movie directors and artists, Dashun Wang would have written about Shakespeare. James Shapiro did exactly that. And he paid close attention to the circumstances that resulted in his 1606 renaissance. "Once you start seeing those plays are really bunched, you start asking: Well, what accounts for a lot of plays in a very short period of time?" he says.

There is another way of asking this question: Why did Shakespeare have the hot hand?

Shakespeare's creative awakening in 1606 came immediately after he'd temporarily disappeared. He'd gone silent as England went through a national transformation under King James. The world as he knew it was a fundamentally different place. But in addition to all the usual fears and anxieties that come with political upheaval, there was something else on Shakespeare's mind. It was also a plague year.

The horrific disease's latest deadly assault on the London area turned out to be the greatest thing that could've happened to this playwright. Shakespeare was able to take advantage of the circumstances in a way that Dashun Wang would appreciate. He didn't stop writing. He kept going. And his talent was about to collide with the oddest bit of circumstance. What could have killed Shakespeare really did make him stronger.

The plague closed London's playhouses and forced Shakespeare's acting company, the King's Men, to get creative about performances. The players had to hit the road. But as they traveled the English countryside, stopping in rural towns that had not been stricken by the plague, Shakespeare hung back by himself. He was too old to be touring, and he no longer had any interest in acting. He felt that writing was a better use of his time. "This meant that his days were free, for the first time since the early 1590s, to collaborate with other playwrights," Shapiro wrote in his book *The Year of Lear: Shakespeare in 1606.*

Shakespeare also benefited from the plague in a most unsavory manner: the plague killed off his competition. By the early 1600s, boys' theater companies were more popular than adult troupes like Shakespeare's, and the children were getting the best stuff from Shakespeare's rival playwrights because talent attracted talent. It was a cycle that Shakespeare couldn't break. He was writing tragedies for adults. With the possible exception of Queen Anne, the audiences wanted satires starring children. But in the hot summer months when the plague wiped out thousands per week, the people who were most susceptible happened to be the people who'd stolen business from Shakespeare. The King's Men would eventually take back their theater spaces and their playwrights because of this disease that preyed on the young. The plague created the circumstance that enhanced Shakespeare's talent. The world had evolved in his favor. All he had to do was adapt.

And that's when it clicked for Shakespeare. That's when he got hot. That's when *King Lear, Macbeth,* and *Antony and Cleopatra* came rushing out of him.

"Three really extraordinary tragedies," Shapiro says. "I'm always interested in how and why this mysterious thing happens of understanding fully the world that you are in and being able to speak to it and for it." It's often tempting for the scholars to scrutinize certain moments of Shakespeare's career through the lens of his personal life. The issue with that line of research is that they still don't know all that much about it. "We have no idea what he was feeling," Shapiro wrote. "We know a great deal more about how a rodent-borne visitation in 1606 altered the contours of Shakespeare's professional life, transformed and reinvigorated his playing company, hurt the competition, changed the composition of the audiences for whom he would write (and in turn the kinds of plays he could write), and enabled him to collaborate with talented musicians and playwrights."

Shakespeare was never a metronomic writer. He was streaky. He wrote in runs. And this run was dependent on forces beyond his grasp. The plague turned out to be the unlikely opportunity of his lifetime. It was because of the plague that he was able to turn a period of great societal upheaval into something else altogether: his hot-hand period.

Shakespeare was talented enough to enjoy some modicum of success no matter his circumstance. Rebecca Clarke was, too. But he made one success into two and three only because those conditions broke his way. His plays were neither random nor independent. One led to another, which led to yet another, which would hopefully lead to another. Shakespeare's ability to capitalize on his circumstance was a form of power never afforded to a talent like Clarke.

So was it him? Or was it the world around him? It was both.

William Shakespeare changed the world only because the world changed first.

Three

———————

SHUFFLE

"It's just random."

1.

There was something wrong with Spotify. By almost every metric, the company appeared to be a tremendous success, a start-up born in a shabby apartment outside Stockholm that had become one of the world's most popular streaming music services. There were millions of people who opened Spotify on their computers and tapped the Spotify icon on their phones to experience a sort of miracle: the ability to listen to almost any song in a matter of seconds. But still they weren't satisfied. Even when the company was already well on its way to global domination, Spotify kept hearing one complaint from a surprisingly large number of irritated customers. "The users were asking 'Why isn't your shuffling random?'" said one Spotify engineer named Lukáš Poláček. "We responded 'Hey! Our shuffling is random!'"

It went on like this for a while. Spotify's users insisted that shuffle wasn't random. Spotify engineers assured them otherwise. There was even a conspiracy suggesting the algorithms were biased toward certain artists to gain favor with their record labels. The truth was less dramatic but no less interesting. But the shuffle button had become so frustrating that Spotify's users felt the only appropriate response was to accuse the company of failing them and maybe even cheating them along the way.

From the early days of the company, Spotify had used the same algorithm for its shuffle button. It was called the Fisher-Yates shuffle. Named for the statisticians who scribbled three lines of code that could randomize any finite sequence, it was an elegant solution that was still being hailed by engineers nearly a century later. To a specific type of geek, the Fisher-Yates algorithm was the *Mona Lisa*. A good number of those geeks worked for Spotify. But it doesn't take a degree in computer science to understand or appreciate the simple beauty of the Fisher-Yates shuffle.

Here's how it works. Let's say there are nine songs in a playlist. We'll call that the existing sequence. Each song gets assigned a number between one and nine. Pick any number up to nine (call it n) and remove that nth number from this existing sequence to begin a new sequence. Then repeat that process with any n up to eight, seven, six, etc., until there is nothing left in the existing sequence. It will look something like this:

N	Existing Sequence	New Sequence
4	1, 2, 3, 4, 5, 6, 7, 8, 9	4
1	1, 2, 3, 5, 6, 7, 8, 9	4, 1
5	2, 3, 5, 6, 7, 8, 9	4, 1, 7
1	2, 3, 5, 6, 8, 9	4, 1, 7, 2
2	3, 5, 6, 8, 9	4, 1, 7, 2, 5
4	3, 6, 8, 9	4, 1, 7, 2, 5, 9
1	3, 6, 8	4, 1, 7, 2, 5, 9, 3
1	6, 8	4, 1, 7, 2, 5, 9, 3, 6
1	8	4, 1, 7, 2, 5, 9, 3, 6, 8

That certainly looks random, doesn't it? But now imagine there's a family in a car for a road trip. They create a Spotify playlist for the ride. Everyone gets to pick three songs. The dad picks three songs by Billy Joel. The mom picks three songs by the Beatles. The daughter picks three songs by Beyoncé. We'll give each of those songs a number: 1, 4, and 7 for Billy Joel; 2, 5, and 8 for the Beatles; 3, 6, and 9 for Beyoncé.

Now let's go back to our playlist—the one created by a randomness generator—and let's see what that sequence of 4–1–7–2–5–9–3–6–8 sounds like:

Billy Joel
Billy Joel
Billy Joel
The Beatles
The Beatles
Beyoncé
Beyoncé
Beyoncé
The Beatles

Wait! That doesn't look very random now, does it? You know that it is. You made it yourself. It's just not how you've been trained to think about random distribution. This family wants the Beatles, Billy Joel, and Beyoncé to be spaced evenly across the playlist. They don't want three Billy Joel songs in a row. By the time "Let It Be" starts playing, there's a good chance someone in the car will be screaming.

This was Spotify's problem in a nutshell. The users were not psyched about hearing three songs in a row by the same artist. And the problem was not unique to Spotify. It was so universal that a competing business found itself struggling with the exact same problem several years earlier.

The original iPod was a glorious device that gave people the ability to carry portable jukeboxes in their pockets. But not everyone who owned Apple's latest release was pleased. Many of them suspected their iPods were defective—that the shuffle button was broken. Their random music wasn't actually random. "It really is random," Steve Jobs said onstage in 2005 during the first keynote speech after the iPod Shuffle was released. "But sometimes random means you've got two songs from the same artist next to each other."

Steve Jobs was extremely Steve Jobs that day. He wore the black turtleneck and jeans, and he spoke with the intoxicating confidence of someone who had seen the future and was reporting back from a rosy society full of Apple products. But not even the most compelling public speaker alive was able to convince anyone that the iTunes shuffle was truly random. The reason that he was onstage talking about randomness was that Apple was introducing a fresh new feature that day. It was called "Smart Shuffle," and it let iPod users control how often they heard consecutive songs from the same artist. Smart Shuffle made sure there wouldn't be three Billy Joel songs in a row. As he explained how it worked, Jobs couldn't help but laugh at the absurdity.

"Even though people will think it's more random," he said, "it's actually less random."

The issue that once demanded the attention of Steve Jobs was now becoming impossible to ignore in Spotify's hip offices. There was something uniquely Swedish about the company's headquarters. The architects left plenty of room in the office for *fika,* the traditional Swedish coffee and pastry break, and Spotify encouraged employees to mingle with people in other departments. Spotify was trying to increase the chances of accidental encounters and the sort of spontaneous interactions that result in the cross-pollination of ideas—the workplace version of the shuffle function.

Spotify was the kind of company that wouldn't blink at entrusting a tricky problem to someone who didn't technically work for Spotify. Lukáš Poláček was still a student at KTH Royal Institute of Technology in Stockholm and wasn't in Spotify's office every day. He picked his projects and focused only on areas where he thought he could make a difference. "I was just looking for stuff that I wanted to improve," Poláček says. He was still looking for stuff that he wanted to improve when he noticed an internal discussion about the shuffle issue. Poláček was studying theoretical computer science and happened to be working on randomness algorithms. Here was a way to make his expertise in computer science a whole lot less theoretical. Poláček volunteered to help.

His narrow slice of expertise in this matter was the reason that Spotify wanted someone like him on the team in the first place. But a strange number of colleagues were confused when he told them he worked on the shuffle button. "What is there to work on?" they would say. "It's just random!" In that sense, they were right. There *wasn't* much to work on. It *was* just random.

Poláček required one day of work and roughly fifteen lines of code to write the algorithm that saved Spotify. It wasn't a feat of technical engineering that made him beam with pride. The man who would be called "Mr. Shuffle" at Spotify parties simply took different songs by the same artist and distributed them more or less evenly across the playlist. It was his job to make sure that Spotify users never had the burden of hearing three Billy Joel songs in a row. Spotify had ripped a page from the Apple playbook. There was only one way to make their playlists feel more random: make them less random.

Before they decided how random was too random, though, Spotify's engineers had to answer a more pressing question: Why was any of this necessary in the first place? They assigned a team to investigate and put a product manager named Babar Zafar in charge. He determined that Spotify had no choice but to remake the shuffle button as quickly as it could. Any other course of action would have been delaying the inevitable. The real problem was one that no amount of money or engineering talent could solve. It was the fact that human beings are pitiful at understanding randomness. There was something about the way that randomness paralyzed the human mind that was beyond the control of Spotify or Apple or any other billion-dollar company. "Our brain is an excellent pattern-matching device," Zafar said. "It will find patterns where there aren't any."

This powerful machine in our heads is also the reason that we believe in the hot hand even when there is no such thing as the hot hand.

2.

Silicon Valley before it was Silicon Valley was a magical place in the memory of Tom Gilovich. As a child in the 1950s and 1960s, when technology hadn't yet transformed the strip of land that he called home, he climbed the hills in springtime and found himself surrounded by apricots, plums, and cherries, and he wandered outside in winter and shot his BB gun in the expanse that later became Apple headquarters. While he was the first person in his family to attend college, graduating from the University of California at Santa Barbara, Gilovich wasn't made to feel lesser than kids who came from wealth. When he decided to pursue a graduate degree after college, he yearned to get back to his paradise. There were lots of reasons anyone with Gilovich's interests would have chosen Stanford University for graduate school, but the gravitational force tugging at him was the same one that he'd appreciated in his childhood: the people.

The faculty of Stanford's psychology department was the primary reason that Gilovich and so many others like him reached the conclusion that there was no better place on the planet to study why human beings are the way they are. He was also pleased to discover that the people in Stanford's psychology department were pretty similar to those people he'd grown up around except for the fact that, by a very strange confluence of events, they were famous. When he was around these professors, young Tom Gilovich was a roadie to academic rock stars.

The ascent of Stanford's psychology department around this time coincided with the rise of Stanford itself. In the 1950s and 1960s, the university had embarked on a strategy once described as "carefully selecting faculty in carefully selected fields." The visionary behind this plan, Fred Terman, took a special interest in the success of Stanford's psychology department. He was an engineer by training, but his wife was a graduate student in the Stanford psychology department. In fact, she had come to study with a famous Stanford psychologist: his father. As the university's provost in the 1950s, the younger Terman was tasked with bolstering Stanford's academic offerings, and he built the school's reputation atop what he called "steeples of excellence." His theory was that Stanford should bet on a certain number of academic disciplines and lavish them with resources. If he picked correctly, those few departments could lift the entire university. Terman identified some obvious steeples of excellence in the hard sciences: aeronautical engineering, mathematical cryptography, space physics, nuclear and chemical weapons research, that sort of thing. The country was recovering from the atrocities of World War II and preparing for the Cold War, and the government was pleased to throw funding at researchers who could assist the military-industrial complex. There were gobs of money sloshing around for people working in national defense.

The psychology department wasn't the most obvious steeple of excellence. It turned out to be a fantastic investment anyway. According to one survey of top graduate schools, Stanford's psychology department was ranked fifth in the country immediately after Terman became provost. But seven years later, when the American Council on Education repeated the survey, it was first. And when Gilovich arrived in the late 1970s, Stanford felt like the center of the universe. There were so many of the field's leading minds in one building that some bad fish tacos at a faculty lunch would've set the whole field back a few days. The many disciplines of psychology had previously been scattered around the university's campus: social psychology in one place, cognitive psychology in another, developmental psychology in yet another. But those specialties were converging at the same time that Stanford's psychology department moved under one roof. The timing couldn't have been any better. The coming revolution in psychology would combine all these disparate branches, and that fusion of psychological specialties would happen first at Stanford. "We *were* the revolution," says the Stanford psychologist Lee Ross.

By the time Gilovich had arrived, the revolution was basically over. Stanford had won. It was clear why Gilovich picked Stanford. It was less clear why Stanford picked Gilovich—or anyone else for that matter. When someone like him applied to graduate school, the psychologists would look at the person's grades, test scores, and letters of recommendation. "But mainly we wanted to see if the person had ideas," Ross says. "What's your *idea*?" What he meant by an *idea* was a clever little way of illuminating something much bigger. The gatekeepers of Stanford's graduate school for psychology wisely concluded that Gilovich would fit right in. "Tom is really exemplary in terms of having neat ideas and having a nose for phenomena," Ross says.

It was by happy coincidence that the very first course Gilovich took in his very first semester at Stanford was not only taught by those professors but was a class entirely *about* those professors. The seminar was called Meet the Faculty. Ross had devised a better way of introducing graduate students to their intimidating new colleagues than sticking them in a room and forcing them to make awkward small talk. Instead they would discuss a classic paper written by a professor in the Stanford psychology department with the author sitting right next to them. It was like a book club reading *Pride and Prejudice* and inviting Jane Austen to lead the conversation.

But in the first meeting of the course, Ross had a surprise for the graduate students. The people they would be reading that day were not fully tenured professors in Stanford's psychology department. They were a couple of Israelis with funny views of the world. Gilovich had never heard of them or read their papers, and he was briefly disappointed. These guest lecturers on sabbatical at Stanford weren't the people he was expecting.

Why can't we hear from the famous people? he thought.

He forgot about his disappointment shortly after the Israelis opened their mouths.

This is amazing! he thought. *Who are these guys?*

The people leading his seminar that day were Amos Tversky and Daniel Kahneman. They would soon become the most famous people in the history of cognitive psychology.

The first time that Daniel Kahneman laid eyes on Amos Tversky in 1957, he couldn't help but pay attention to the thin, pale man with a red beret perched atop his head. Tversky was on his way from taking the exam required for entrance to the most competitive field of study at Hebrew University of Jerusalem: psychology. There were twenty spots available for hundreds of interested students, Michael Lewis wrote in *The Undoing Project: A Friendship That Changed Our Minds*, his masterful account of Kahneman and Tversky's partnership. What made it so competitive was that if you were young, smart, and Israeli in the immediate aftermath of World War II, chances were that you wanted to understand the vagaries of the human mind.

Amos Tversky was young, smart, and certainly Israeli. He was a paratrooper in the army, which explained the red beret, and he'd been promoted to platoon commander when his military unit found itself performing a fateful training exercise. The goal of the drill was to blow a hole through a barbed-wire fence, and the plan was to place a grenade near the fence, light the fuse, and retreat. But it did not go to plan: Tversky's soldier failed to retreat. He froze. They would have been the last seconds of his time on earth if not for his platoon commander. With the clock ticking and his superior ordering him to halt, Tversky sprinted to his soldier, dragged him away from the torpedo, and covered him with his own body right before it detonated. He saved the soldier's life by a matter of seconds. "Those who have been soldiers will recognize this act as one of almost unbelievable presence of mind and bravery," Kahneman said many years later, long after the Israeli military had awarded Tversky its highest honor. By the time he was twenty years old, Tversky was a decorated hero.

It was not the last accolade that he'd win. One of the prizes that came a long time after his military commendation was the MacArthur Foundation Fellowship. It's commonly known as the "genius" grant. But the most remarkable part about Tversky officially being recognized as a genius is that it was redundant: everyone who knew him already knew that he was a genius. The joke among them was that he even had an intelligence test named after him. It measured how long it took you to realize that Amos Tversky was smarter than you. If you measured their intelligence by the Tversky test, the people at Stanford were phenomenally bright. In this building with the greatest collection of psychological brainpower assembled in one place, they realized they were all trying to keep pace with this visiting professor.

Kahneman was among the first people to have recognized Tversky's brilliance. Kahneman was raised in Paris, fled to the South of France when the Nazis invaded, and moved to Israel after the war. He studied psychology at Hebrew University in Jerusalem and moved to the United States for graduate school in 1958 after his mandatory service in the Israeli army. About a decade later—after getting his Ph.D. at the University of California at Berkeley, teaching at Hebrew University, taking a sabbatical at the University of Michigan, and continuing his research at Harvard University—Kahneman returned to Israel.

It was in the spring of 1968 when he crossed paths again with that young paratrooper in the red beret who had been waiting to be admitted to the Hebrew University undergraduate psychology program about a decade earlier. They had run into each other when they overlapped at American universities. But when Kahneman went to Harvard, Tversky went home. He was the newest faculty member of that very same Hebrew University. By the time Kahneman invited him to speak with a graduate seminar about judgment, Kahneman's field of expertise, he was coming around to a judgment of his own. He was beginning to suspect that Tversky was brilliant. They went for lunch after class and fell into an easy conversation. It was the first in a series of conversations about ideas that lasted for decades. What began that day over lunch was their decades-long study of judgment and decision-making.

One of their first collaborative efforts was a survey they distributed at a 1969 meeting of the American Psychological Association. Kahneman and Tversky polled the highly trained scientists they called their colleagues and found something that surprised them: their intuitions about randomness were leading them astray. It wasn't merely ordinary people who had trouble understanding randomness. It was also the people whose jobs required them and specifically trained them to understand randomness. It was everyone—including Kahneman and Tversky.

The conclusion they drew from their questionnaire was classic Kahneman and Tversky before there was such a thing as classic Kahneman and Tversky. Reading it now is like hearing grainy recordings from grungy Hamburg clubs and recognizing bits and pieces of the band that would eventually become the Beatles. Even in their first coauthored paper, Kahneman and Tversky didn't write like typical psychologists. They defied the traditional format of introduction, methods, results, and discussion. Instead they weaved their empirical results into a larger narrative that described and then analyzed the phenomenon they were observing. While they had the audacity to be readable, they were never arrogant about their work. They couldn't be. They laughed at every bias and error they uncovered because they had fallen for the same biases and errors themselves. In fact, that was the most amusing thing about it.

When it came time to publish their first paper in *Psychological Bulletin,* they flipped a coin to decide whose name would be listed first, since their contributions were split down the middle and embracing randomness seemed like the only fair way to settle the issue of lead authorship. Their first paper was by Tversky and Kahneman. Their next paper was by Kahneman and Tversky. They alternated the order of their names for every paper that came afterward.

The breakthrough of their collaboration was a 1974 paper (by Tversky and Kahneman) in the prestigious journal *Science* that came to be known simply as "the *Science* paper." It took a year to write. They worked in pencil and cycled through dozens of drafts. Every word mattered. On a productive day, they wrote one sentence. This was their process for making poetry of technical academic papers. "When the initial idea is good—and with Amos it almost always was—you end up with something that cannot be improved further," Kahneman later wrote.

The *Science* paper had so many good ideas and was so carefully written that it's hard to imagine how it could have been improved further. Their basic thesis was that humans rely on certain rules for decision-making. Those rules are generally useful, except when they're not. "Sometimes they lead to severe and systematic errors," the two Israelis wrote. This paper dealt with those painfully human errors—the cognitive biases and illusions that fool us. They identified the differences between how we should make decisions and how we *actually* make decisions.

One of the biases they explored was something they called the "law of small numbers." The law of large numbers suggests that large data sets are unlikely to be skewed by outliers. But the law of small numbers stipulates that people infer too much from too little—that we tend to have "exaggerated confidence in the validity of conclusions based on small samples," as they put it. The belief that the results of small samples hold for large samples could be erroneous, in other words, because the outliers that distorted small samples disappeared in large samples. But there were few people who were above the law of small numbers. Kahneman and Tversky's research showed that even some experts were prone to this bias in their fields of expertise. They spotted the law of small numbers everywhere from academic psychology to the Israeli military, and they would have found it anywhere else they bothered to look.

"Consider a hypothetical scientist who lives by the law of small numbers," they wrote. "Our scientist could be a meteorologist, a pharmacologist, or perhaps a psychologist."

That last example was cheeky, but it also served a purpose. It was a powerful reminder that Kahneman and Tversky were not studying meteorologists or pharmacologists. Their work was so delightful because they were studying their colleagues in psychology. In a way they were studying themselves. "They chose concrete examples," Ross recalls. "They chose materials to operationalize their ideas."

Tom Gilovich had been unfamiliar with Kahneman and Tversky when he enrolled in Stanford's psychology department, but the first thing he read in his first course on that first day of graduate school was the *Science* paper. He was instantly enraptured. Gilovich couldn't stop thinking about these men who told him that he'd been looking at the world the wrong way.

Tversky would soon become a professor at Stanford, and Gilovich would take his seminar in judgment and decision-making. Tversky recommended to him a paper about randomness that showed

Gilovich how we assign sense to nonsense and detect order where there is none. Gilovich began to wonder about the other ways our minds can play tricks on us.

We see a few spots on the surface of the moon and think the cosmos are sending a message. We hear a few songs from the same artist and think that Spotify's playlists are screwy. We look at successful investors and confuse luck for skill. The most famous example of this illusion was the Germans dropping bombs on Great Britain during World War II and Londoners convincing themselves there had to be some rhyme or reason for where they landed. There wasn't. It was random.

"People see patterns where there are none," Tversky once said, "and they invent causes to explain them."

If this paper was right about people seeing patterns in randomness, and he was pretty sure it was, if only because Tversky said so, Gilovich suspected they would stumble upon it everywhere. There was one place in particular where he wanted to look for this phenomenon: basketball.

He didn't have to sell his academic mentor on the notion that basketball could be fertile ground. Amos Tversky *loved* basketball. It was one of the few topics—along with Israeli politics, the big bang, and astrophysics—that enthralled him as much as his own field of study. Of all the millions of people who have devoted huge chunks of their brains to basketball, surely Tversky was the smartest. Gilovich wasn't too far behind him. He was a pretty good player, too. There was one year when the Stanford psychology department's basketball team (and some ringers of Gilovich's choosing) made it all the way to the finals of Stanford's intramurals tournament. They lost on a buzzer-beater to the Stanford football team. The star of the Stanford football team's basketball team was none other than the future National Football League quarterback John Elway.

Gilovich was far too shy to declare a topic like randomness in basketball worthy of Tversky's time and attention. Tversky was a towering intellectual titan who could have chosen any subject and probably would have revealed something interesting about the human mind. "Amos had simply perfect taste in choosing problems," Kahneman once said, "and he never wasted much time on anything that was not destined to matter."

So when Gilovich visited Tversky's office a few months after his first day of graduate school, he mentioned only that randomness might be misperceived in basketball. He figured our poor sense of intuition could make us see things that weren't there. His theory was that basketball players and fans might have an exaggerated sense of the hot hand. It was an insight that touched on many of the underlying points of the *Science* paper, and Tversky didn't need much persuading that the hot hand might be a myth.

In typical Amos Tversky fashion, he'd already toyed with the idea himself. He'd even tried compiling the data by recording Boston Celtics games and tracking how players shot when it seemed like they were hot. But he kept running into the problem that would vex other hot-hand researchers: there was no way to perform a rigorous analysis of the hot hand without coding thousands of hours of basketball. Not even Tversky cared *that* much about basketball. But Gilovich had read about a compulsive statistician with the Philadelphia 76ers who did, and he hoped this professional dork would be obsessive enough to have the numbers they needed for a proper study of the hot hand. Gilovich wasn't suggesting that study, though. Really he was just inventing an excuse to be around Tversky. If Tversky had shooed him away that day, Gilovich would have forgotten about the hot hand. Tversky could have told him that he was wrong, that *of course* the hot hand was real, and Gilovich would have believed him.

"I wouldn't have had enough confidence to say I'm right and Amos is wrong," he says.

Amos Tversky did not think Tom Gilovich was wrong. His experience watching all that basketball made him sympathetic to the notion that this graduate student was onto something. But he had a slightly different hypothesis. Tversky didn't think the hot hand was exaggerated. He thought it didn't exist. And he thought investigating that possibility was an idea that had potential.

"It so beautifully exemplifies the particular strengths of the way that Amos worked," Ross says. "What is the phenomenon? 'That people have erroneous notions about randomness and think they're seeing non-randomness when they're looking at chance distributions.'"

"That's not a vivid, memorable statement," he continues. "At least not compared to: 'Basketball fans see hot hands everywhere, but the statistical analysis doesn't bear that out.'"

That's what Amos Tversky loved about Tom Gilovich's suggestion to look at the hot hand in basketball. It was how an idea could be operationalized. Because the hot hand in basketball wasn't really about basketball. It was about behavior. That was the reason Gilovich soon received a short message that would change the course of his career: "Amos wants to see you."

3.

A cognitive psychologist studies the hot hand and thinks: *Why are we stupid?* An evolutionary psychologist studies the hot hand and thinks: *What if we're not actually being stupid?*

"I'm hesitant to think of the hot-hand bias as something that originates in North American basketball," says Andreas Wilke. "I find that personally very unsatisfactory. And not only because I'm an evolutionary psychologist. And I'm German. And I think about soccer more than basketball."

As a professor at Clarkson University in upstate New York, Wilke wasn't interested in whether the hot hand was real as much as he was interested in the reason that human beings had never been able to shake this bias. "Could it be that this bias is not a bias, per se, but a carryover from evolutionary times?" he says.

Wilke thought it wasn't a bug of the human brain. It was a feature. It was our biological way of recognizing patterns and processing information in order to help us survive. This was something Darwinian that had been ingrained over the course of twenty-five million years. It turns out evolution favored those who followed the hot hand.

To understand how the hot hand could be a cognitive adaptation, not a cognitive error, it helps to understand the original purpose of detecting patterns. As an evolutionary psychologist, Wilke was used to looking backward for clues about modern behavior. When he thought about the hot hand and why humans believed it, his best guess was that our ancestors relied on the hot hand to forage. The resources they needed in their daily lives weren't randomly distributed nor proportionally dispersed. They were clumped. Clumps are the norm in nature. There are clumps of food, clumps of water, clumps of shelter, clumps of people, and clumps of information. There are clumps on clumps on clumps. There were patterns to be found and existential rewards for the early humans who found them.

The world has changed since then. It's not so clumpy anymore. But *we* haven't. We're still looking for patterns even when our environments are effectively random.

At least that was his theory. It was a hard one for Wilke to test. It would have been difficult for him to turn back the clock millions of years and recruit primitive societies for his study. Wilke had neither the money nor the technical expertise to build a functional time machine.

But he didn't have to. Wilke collaborated with H. Clark Barrett, a biological anthropologist who simply hopped on a plane instead. Barrett was used to conducting field studies far away from the laboratory with subjects who weren't college freshmen in need of Psych 101 credits. And this particular study used perhaps the most dissimilar people I've encountered in scientific research: indigenous Shuar hunter-gatherers in a rural Amazonian village and students on the dreamy campus of UCLA. These foragers in a natural environment were the closest that Wilke could get to probing his ancestors. It was a clever solution to the inherent problem of being an evolutionary psychologist.

Barrett made the journey to Ecuador by riding buses that took him to more buses that took him to a truck that took him down a dirt road that took him to his test subjects. He braved the mosquito bites long enough to show them a piece of technology that hadn't yet reached their remote village: his laptop. On the screen was a series of tests that would measure how much they believed in the hot hand. The Shuar foragers and the UCLA students played the same computer game. The task was mind-blowing for one group and banal for the other. Barrett showed them a sequence of one hundred hits and misses one at a time. They had to guess whether the next outcome would be a hit or a miss and were given cash prizes for every correct prediction. (The UCLA students were offered a more lucrative exchange rate than the Shuar foragers.)

The game was designed so there would always be a 50 percent chance of a hit or a miss. The foragers and the undergrads were basically calling heads or tails on coin tosses. For one of the experiments, that's exactly what they were doing. The object in the game was a coin, and they made their bets before it flipped. For the other experiment, however, the coin became a tree. The bet was whether the next tree would be empty or full of fruit.

Wilke and Barrett noticed a difference between the UCLA students and the Shuar foragers when they tallied the coin-toss results. UCLA students realized the coin flips were random and behaved accordingly, but the Shuar foragers made their guesses of heads or tails in streaks. Coin flips were just like the fruit trees and other natural resources in their minds. They were clumped.

The researchers saw no difference between the two groups, however, when they looked at the fruit tree results. The UCLA students behaved the same way as the Shuar foragers. The natural resource was deeply unnatural to these college kids living in a city. They were more likely to bet on a hit when there was fruit on the previous tree, and they were more likely to bet on a miss when the previous tree was empty. Hits followed hits. Misses followed misses. What the Shuar foragers and UCLA students had in common—maybe the only thing they had in common—was their shared belief in the hot hand.

The hot hand was not simply a "glitch in the system," Wilke and Barrett wrote, or some "byproduct of Western industrialized culture." It existed because there were real evolutionary benefits to its existence. "The hot hand," they concluded, "is a pervasive feature of human thinking."

But what about nonhumans?

That was the question on Ben Hayden's mind as he drove to meet Wilke from his teaching post at the University of Rochester. Hayden had long been intrigued by the notion of the hot hand, and not only because he was a cognitive neuroscientist. In his last year of graduate school, Hayden played basketball almost every day. "I'm big and slow," he says, "but when I'm playing a lot, I'm actually a pretty decent shooter." There were some days that year when he was better than pretty decent. On those days, he was a dorky Stephen Curry. Hayden noticed that being in the zone had a tendency of leaking into other parts of his life. If he had a good day in the lab, he would have a good day on the basketball court. He used to play squash against someone who swore by the hot-hand effect. Hayden told him that he was nuts. To prove his point, his squash partner started tracking their results. "The data was right," Hayden says. "The data supported him."

Like any good scientist, Hayden was loyal to the data. The data was the reason that he was a scientist in the first place. He was an undergraduate chemistry major on his way to graduate school to study what he called "molecular stuff" when he finally took a humanities course in the second semester of his senior year to fulfill his graduation requirements. The class was called Philosophy of the Mind. He found it interesting—and infuriating. What frustrated him was the total lack of objective evidence in their discussions of free will. "You guys could answer these questions so easily if you could stick electrodes into the brain," he remembers thinking. "You could solve these stupid philosophical debates that people have been having since Greek times with a little bit of data." Though he would go on to earn a Ph.D. in molecular and cell biology, Hayden was so invigorated by the lessons of his one philosophy class that he focused on the psychology of decision-making, specifically concentrating on animals and cognition. The sweet spot of his scholarly interests was how monkeys and humans deal with uncertainty. That was the research Wilke wanted him to discuss when he invited him to campus for a talk, and they were having lunch afterward when the conversation turned to a topic that encapsulated all their fields of study: the hot hand.

"We should test this in monkeys!" Hayden said.

The reason people like Ben Hayden and Andreas Wilke like to study monkeys is that it helps them find out how far back in our evolutionary heritage these biases existed. "Are they innate?" Hayden says. "Or did they happen within our lifetimes?"

To answer this question, Hayden picked three of the juvenile rhesus macaques from the University of Rochester's colony of monkeys. He already knew that humans were prone to seeing patterns in randomness. He wanted to know if our last common ancestors were, too. He really wanted to know if something that makes us stupid was at one time smart.

But first he had to make his monkeys happy. And what made them happy were the same pleasure centers that fired in the minds of humans. "You have huge incentives to make them happy because they're going to give you more data and better data," he says. "But you don't just give them food. You make them forage for it. You're constantly going to the hardware store to build them puzzle contraptions so they have to forage."

The rhesus monkeys were seated in ergonomically designed chairs in a testing room of Hayden's lab that was painted black, electrically insulated, and completely silent to limit distractions. The experiment began. They stared at the computer monitor. There were two side-by-side photographs of nature. The monkeys were instructed to choose one of the photos by moving their eyes to the left or to the right. If the monkey selected the proper target, which was the photograph that Hayden and the other humans wanted the monkey to pick, the monkey would be rewarded with shots of water and cherry juice.

The monkeys made thousands of selections as the humans observed from another room. This was like watching a traffic jam in a parking garage. "It's really thrilling for five minutes," Hayden says, "and really boring for three hours." But there was something different about each day of the experiment, and it was why this study of monkeys is relevant to humans. The photo that earned the monkeys more juice changed sides. Some days the alternation rate was 10 percent. Some days the alternation rate was 90 percent. But it was the same percentage for a full day. That should have given each monkey plenty of time to adapt and make decisions that maximized its juice.

The goal of the monkeys was to get as much juice as possible. The goal of the humans was to see if they picked clumpy resources over dispersed ones.

When the photo remained on the same side more than 50 percent of the time, the optimal strategy was for the monkey to keep staring at that side. It was easy for the monkeys to wrap their monkey brains around this strategy. What they were really doing was foraging clumpy resources. The three monkeys sipped their juice 90 percent, 87 percent, and 84 percent of the times when the photos stayed on the same half of the computer screen. They had no trouble when the game incentivized them to embrace the hot hand.

But when believing in the hot hand was suboptimal, they failed miserably. This is when Ben Hayden's monkeys went thirsty. The monkey with the 90 percent score dropped to 71 percent. The monkey with an 84 percent fell to 70 percent. And the monkey with an 87 percent must have been parched since its score plunged to 33 percent.

The monkeys couldn't bring themselves to pick the dispersed resources even when they had a strong, tasty incentive. And the humans watching the monkeys were reminded that they weren't so different after all. We both have a strong belief in the hot hand. "They have that weird, freaky bias that I do," Hayden says.

4.

Tom Gilovich and his research assistants walked into a gym one day in 1982 to interview a collection of very tall, very sweaty lab rats: the Philadelphia 76ers.

The professional basketball players were ushered over to the psychologists one day after practice for a field experiment that had been engineered by Harvey Pollack. Like his visitor from Stanford, Pollack was also obsessed with basketball and intrigued by the notion of the hot hand, but he wasn't a scholar. He was the statistician for the 76ers. Pollack's devotion to statistics was so unusual for the time that he was given the nickname "SuperStat." He was exactly the man Gilovich was looking for. Pollack had the information that Gilovich and his assistants needed. In fact he was the *only* person with that information, considering he was the only person in the NBA who bothered keeping track of shots in the order they were attempted. Gilovich had called him out of the blue asking if he could help with a study of the hot hand, and Pollack was pleased to send photocopies of his chicken scratch to these researchers at Stanford who promised to make sense of it. Amos Tversky, Tom Gilovich, and Robert Vallone suddenly had the 76ers' shooting records and sequences for their home games in the 1981 season.

The useful thing about having lab rats who were NBA players instead of vermin was that they could talk back when asked questions. Gilovich and his research assistants had lots of questions for the star players like Julius Erving and Darryl Dawkins. But mostly they wanted to know if the players believed there was such a thing as the hot hand.

Did they ever feel like they couldn't miss a shot after making a few shots in a row? The 76ers said yes. Did they think they had a better chance of making their next shot after making a few shots in a row? Yes. Did they take more shots after making a few shots in a row? Yes. Did they think it was the right strategy to pass the ball to the player with the hot hand? The 76ers said yes emphatically.

This was exactly what Gilovich expected them to say. In fact he'd already surveyed one hundred basketball fans with the same questions. They had provided the same answers as the Sixers. They were absolutely convinced the hot hand existed. But that wasn't all. Gilovich then presented these fans with a hypothetical player who made 50 percent of his shots. What was that player's shooting percentages, he asked them to estimate, after making a shot and missing a shot? The participants in this survey were smart basketball fans from Stanford and Cornell, but their responses defied

probabilistic reason. They estimated that a 50 percent shooter
became a 61 percent shooter after *making* a shot but a 42 percent
shooter after *missing* a shot. That didn't seem to make any mathematical sense. Their belief in the hot hand was so strong that it
appeared to break their brains.

Their convictions had all the classic signs of a cognitive bias if
the evidence proved contrary to their intuitions. As it turned out,
Gilovich, Vallone, and Tversky already had such evidence. They
looked at the sequences of Sixers shots that Pollack had given them
and calculated each player's shooting percentage immediately after
making a shot and missing a shot. If there were such a thing as the
hot hand, the probabilities would be higher after made shots. They
weren't. The players they analyzed were more likely to make a shot
after missing a shot than they were after making one. When they
thought they were hot, they were not. On those glorious occasions
when Sixers players made three shots in a row—when Tim Kitzrow's
synthesized voice would've said they were *on fire*—they shot worse
than they did when they had missed a few shots in a row.

Gilovich, Vallone, and Tversky failed to detect any signs of streak
shooting among the Philadelphia 76ers. But it wasn't only the Sixers.
They also secured the free-throw shooting numbers of the Boston
Celtics. There was no hint of the hot hand there, either. In order
to eliminate the little doubt that was left, the researchers arranged
for an experiment in a more controlled environment, a basketball
court they refashioned as their laboratory. They invited twenty-six
players from Cornell's men's and women's teams to the gym one day
and determined the spot on the court where they made roughly half
their shots when no one was guarding them—where their basketball
shots were coin flips. Each player attempted one hundred shots from
those spots and made a prediction before every one about whether
it would go in.

Since they were paid for accurate predictions, they had a clear incentive to be right. The subjects were being asked to bet on themselves—their basketball ability as much as their intuition. Before every shot, they had to make a wager: five cents for a make and four cents for a miss, or two cents for a make and one cent for a miss. The hotter they felt, the bigger they bet. But they weren't the only gamblers in this experiment. Gilovich also collected bets from their rebounders. He was testing whether observers were any more skilled at predicting the outcome of a shot.

The results plunged the hot hand into a tub of ice. The shooters were predictably horrible at guessing which shots they would make and which shots they would miss, even when they felt there was no doubt they had a hot hand. Their rebounders weren't any better. The shooters and rebounders were also more likely to increase their bets on a hit simply because the previous shot had been a hit. They were biased toward the hot hand, and they let those biases seep into their bets.

Gilovich, Vallone, and Tversky had now looked at thousands of shots from the 76ers, the Celtics, and Cornell's basketball teams. They had seen nothing to suggest that basketball players were right to believe in the hot hand. To understand *why* they believed it, though, the researchers had to run one last experiment. They went back to the one hundred students they polled about the hot hand and showed them a series of runs that looked something like the text messages of a teen couple:

XOXOXOOOXXOXOXOOXXXOX

The students looked at six of these sequences. They were not too dissimilar from the sequences that Shuar foragers had been shown. There were twenty-one Xs and Os in all—always eleven Xs and ten Os—but the alternation rate varied. Some of them were more XXXOOO than XOXOXO. The students were instructed to study the Xs and Os and guess whether a sequence represented "chance" shooting or "streak" shooting. Chance shooting is what we think of as randomness. Streak shooting is how we think of skill.

Here's a visual representation of how they responded:

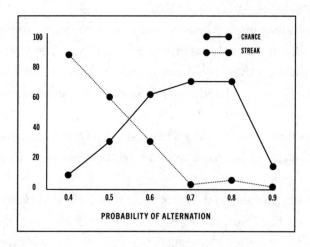

The runs with a low alternation rate are on the left part of the chart. The students identified these runs that look like XXXOOO as streak shooting. The runs on the right half of the chart had a higher alternation rate. The students believed that XOXOXO represented chance shooting. They predictably attributed the clumps to skill.

Gilovich, Vallone, and Tversky focused their attention on the two dots above the 0.5. That's when the alternation rate was a perfectly random 50 percent, and the odds of an X becoming an O were roughly a coin flip. This was the purest distillation of chance shooting. Except that's not how a majority of the students viewed it: 62 percent interpreted chance shooting as streak shooting. They looked at perfect randomness and saw the hot hand.

Why? The underlying explanation went back to the *Science* paper in the Meet the Faculty course that Tom Gilovich had attended and the randomness study that Amos Tversky had given him in his judgment and decision-making seminar. People expected sequences that were random in the long run to be random in the short run, too. But they're not. And maybe the most remarkable thing about this was that experts were more susceptible to the bias than anybody. The experts in this particular experiment were the basketball players. They remembered all the times *they* were hot, and they refused to buy the notion that events seared in their memories were random. As he often did, Tversky said it best: "Very often the search for explanation in human affairs is a rejection of randomness."

Gilovich, Vallone, and Tversky already had the neat ideas that human intuition could be wrong, that people systematically misperceive randomness, and that such biases lead us in all kinds of funky directions. Now they had proof of it, too. That meant it was time to write.

The scholarly paper they published several years later was nothing short of a sensation. The ruckus that ensued was almost entirely because of what they discovered: nothing. But it was the right sort of nothing. There were three major findings of their study. The first was that there is no such thing as a hot hand in basketball. The second was that humans fundamentally overestimate streakiness and have a wildly exaggerated sense of what it means to be hot. The third was that people have a nasty habit of looking at randomness and seeing patterns. "The present data demonstrate the operation of a powerful and widely shared cognitive illusion," Gilovich, Vallone, and Tversky declared, concluding, "Thus the belief in the 'hot hand' is not just erroneous, it could also be costly."

Their paper was so unbelievable that most people simply refused to believe it. It was work that had those costly implications for al-most every industry. It was a window into the human mind. It was contrarian, accessible, and concise. It was written in language that even basketball coaches understood. It was basically everything you could ever want in a paper. But it was promptly rejected for publication by *Science*.

Gilovich, Vallone, and Tversky had stormed the castle of rational thought with their pitchforks and produced a study that was almost too counterintuitive for its own good. If that was how those in charge of *Science* had responded to their paper, you can only imagine how surly basketball coaches reacted once it was published in *Cognitive Psychology*.

But you don't have to imagine it. One lucky reporter actually had the pleasure of sharing the results of their study with the legendary Boston Celtics coach Red Auerbach. Auerbach was not swayed by Tversky's rigorous academic work nor his claim that the hot hand was a figment of the imagination. "Who is this guy?" he sneered. "So he makes a study. I couldn't care less." At this point I like to imagine Auerbach extinguishing his cigar on his desk and storming out of his office in disgust.

The backlash from the basketball establishment was rich. There was nothing that could have delighted Tversky more. The experts with an incentive to correct their error were refusing to do it. In fact they claimed it wasn't an error. In his judgment and decision-making seminar, the same one that Gilovich took when he was a graduate student, Tversky saved his seminal work about the hot hand in bas-ketball until the very end. He liked to close his legendary course with a lecture on the hot hand—including the quote from Red Au-erbach. He relished in telling the story of how Auerbach, unwilling to believe the data, unwittingly proved the point of their paper. The extent to which people thought the hot hand was real didn't make the hot hand any more real. "There are plenty of excellent reasons why the hot hand could exist," Tversky liked to say. "The only trouble is it doesn't."

His work on the hot hand would be one of the last lectures that he gave to his class in the spring of 1996. The great Amos Tversky died of cancer after the end of the semester. He was fifty-nine.

Only a few months earlier, the Swedish committee responsible for awarding the Nobel Memorial Prize in Economic Sciences had secretly amended its rules and regulations, which were things that Swedish committees armed with Nobel Prizes cared very much about. What changed was the very definition of economics. Anyone who advanced the broader field of social science—sociology, political science, and especially psychology—would now be eligible for this Nobel.

By the time they could give Amos Tversky the prize, it was already too late. The Nobel Prize is not awarded posthumously. The first non-economist to win the Nobel in economics would be Daniel Kahneman instead. That he would have shared the laurels with his intellectual partner was such a certainty that Tversky was cited in the Nobel Prize citation given to Kahneman, the first and only time a person other than a Nobelist was mentioned in a citation, and Kahneman's eulogy for Tversky was appended to the official autobiography required of Nobel laureates. There has never been anybody who so obviously won a Nobel Prize without actually winning the Nobel Prize.

But he never really cared about prizes anyway. He cared about ideas. The staggering canon and countless admirers he left behind would keep his ideas alive.

Amos Tversky was beloved. He had the enviable habit of making friends who adored him and idolized his work even if they were intimidated by him. The people who knew him best were the ones who came to realize that it was just about impossible to win an argument with him. It wasn't only Tom Gilovich who didn't have the confidence to believe he was right and Tversky was wrong. It was pretty much everybody who interacted with him. They felt that telling Amos Tversky he was wrong was a bit like telling Stephen Curry how to shoot.

But in his lifetime there was one topic that not even he could get these people to change their minds about. He once told a friend that he'd never encountered more resistance than when he wrote about the elusive, provocative, devilishly entertaining idea known as the hot hand.

"I've been in an endless number of arguments," Tversky said. "I've won them all, and yet I didn't convince a soul."

Four

BET THE FARM

"Principles over patterns."

1.

James Naismith was raised on a farm in rural Canada so far away from civilization that going to school meant walking five miles through the woods. He eventually started taking the easier route and stopped going altogether. The young Naismith chopped trees, split logs, and did all the other things lumberjacks do until he'd done them for a few years and felt it was finally time to get serious about his studies. Once he began walking the five miles through the woods again, he was surprised to learn that he actually liked school enough to enroll in college, where Naismith spent so much time with his books that upperclassmen came to his room one day with some advice: "You spend *too* much time with your books."

Naismith happened to be walking past football practice not long afterward. He wasn't expecting to play and certainly wasn't expecting his life to shift off the course he'd carefully plotted as a strict Christian in theology school preparing to be a minister. But when one of the football players broke his nose, Naismith suddenly found himself on the field taking his place. He was uncomfortable with how much he enjoyed it. He thought he shouldn't be wasting his time with sports, least of all football, which the deeply faithful considered to be a tool of the devil. Now concerned about his embrace of a satanic activity, his friends huddled one day to pray for his soul. But the truth was that Naismith didn't have to choose between his sports and his studies. He could have both. "I thought there might be other effective ways of doing good besides preaching," he later wrote. He was brave enough to share that idea with someone who didn't bother praying for his soul, and Naismith learned of a school in Springfield, Massachusetts, developing young men around that hybrid concept. "I made up my mind," he said. "I would drop the ministry and go into this other work."

The first person he met in his new job was a beloved dean named Luther Gulick. A striking man with red hair and blue eyes, Gulick was in charge of physical education at the Springfield International YMCA Training School. He was also in a pickle. The YMCA offered football in the fall and baseball in the spring, but there was no sport to keep the students busy in the winter. This was Gulick's dilemma. He needed an entertaining game that could be played indoors.

His frustration boiled over at a meeting in the fall of 1891. Gulick had undertaken an extensive investigation into the nature of sports, but he told Naismith and the other faculty members that it had fizzled. The only conclusion he could draw was that every game owed some debt to another game—that all new games were thinly disguised versions of older games. What seems pretty obvious was a profound insight to Naismith. He mused that he could invent a game simply by combining elements of games that already existed. Gulick dared him to actually do it. Naismith initially responded by doing absolutely nothing. But no matter how much he procrastinated, no matter how many excuses he made, no matter how little time he claimed to have, his boss refused to budge. Winter was coming. The students were getting restless. Gulick pulled Naismith aside. "Now would be a good time for you to work on that new game that you said could be invented," he said.

When he finally got to work, it didn't go well. In fact it went terribly. "How I hated the thought of going back to the group and admitting that, after all my theories, I too had failed," Naismith recalled. "It was worse than losing a game." It was only in that moment of desperation that he realized where he'd gone wrong in thinking about this hypothetical sport. Naismith had been directing his energy toward making small tweaks to football, soccer, and lacrosse. But that's all they were: football, soccer, and lacrosse with small tweaks. It was like removing the bun from a hamburger and calling the patty filet mignon.

He sat down at his desk one night and tried to figure out the flaw in his logic. He'd been too focused on the specifics of individual games. But what about games in general? The more he thought about it, the more he realized the sports he liked had lots in common. They had a ball. They had a goal. They had an objective of moving that ball toward that goal. And that's when Naismith had an idea.

His ideal sport actually sounded like rugby, he thought, but rugby couldn't be played indoors because there was tackling, and there was tackling because players could run with the ball. Naismith shifted in his chair. If the players couldn't run with the ball, there wouldn't be a need for tackling. If there wasn't a need for tackling, then a sport like rugby could be played indoors. He snapped his fingers and shouted, "I've got it!" There was no one else in the room.

After his eureka moment, the whole sport came pouring out of him. He played the first game in his head as he fell asleep that night. He skipped to his office the next morning. He picked out a ball and asked the YMCA's superintendent if there were any old boxes lying around the building. "I have two old peach baskets down in the store room, if they'll do you any good," he said.

They would do just fine. Naismith found a hammer and nailed

those peach baskets to the gymnasium wall. It was almost time for class. He reached for his notepad, scratched out thirteen rules for his game, and handed them to his stenographer for typing. Naismith hung the rules inside the gym seconds before his students arrived.

There was only one more thing Naismith's new game needed: a proper name. His class had already been playing his sport for several weeks when a student approached him with two suggestions. The first was "Naismith ball." Naismith laughed in his face. So the first idea was a dud. But after his second idea, Naismith smiled. The second idea was a winner.

"Why not call it basketball?"

2.

David Booth grew up in Lawrence, Kansas, on a street called Naismith Drive. It was his birthright to love that sport called basketball.

His childhood home was a half mile from the University of Kansas's arena, and he often walked over to watch basketball practice after school. But as much as Booth loved basketball, basketball did not love him back. He played as much as any boy in his neighborhood, but he was not destined for a career in the NBA. He threw himself into the study of economics instead. Booth was as captivated by the underpinnings of economics as he was by the intricacies of his favorite sport, and he stayed at the University of Kansas for an extra year to get his master's degree, at which point he'd been in Kansas long enough to know that it was time for him to leave Kansas. He also knew where he wanted to go.

When he was in graduate school, Booth had taken a finance class and read a dissertation called "The Behavior of Stock-Market Prices." He was hooked from the first sentence. "For many years the following question has been a source of continuing controversy in both academic and business circles," the paper began. "To what extent can the past history of a common stock's price be used to make meaningful predictions concerning the future price of the stock?" The thesis went on this way for seventy-one pages of dense economic theory. Booth couldn't put it down. It was just about the most compelling thing he'd ever read. The idea that markets were efficient, that prices reflect everything there is to know about a given asset, made a whole lot more sense to Booth than anything else he'd been told about the way markets worked, which was that investors could exploit information gaps and reliably buy low and sell high. Booth told his finance professor how much he admired this paper and found out that it had been written by a young guy named Eugene Fama who was teaching at the University of Chicago. Chicago's economics department of the 1960s was Stanford's psychology department of the 1970s. "Chicago's the place," his professor said. "That's where everything is happening." David Booth piled everything he owned into his convertible and drove nine hours from Naismith Drive to Chicago.

It was an exhilarating time to be an economics geek at the geekiest place on the planet for economists. As a Ph.D. student at Chicago, Booth witnessed the dawn of modern finance. In addition to everything he learned in the classroom, he also learned that he wasn't the only person who thought Fama was onto something. The grandson of Sicilian immigrants and son of a truck driver in the Boston suburbs, Fama went to a Catholic high school and, like Tom Gilovich, became the first person in his family to attend college. Fama majored in Romance languages and envisioned a future teaching high school and coaching sports. Only when he couldn't stomach reading another page of Voltaire did he bother taking an economics class. It wasn't long before he'd abandoned his dreams of coaching high school sports and talked his way into a scholarship at the University of Chicago. The school's dean had no record of his application when he called, but he'd offered him a spot in the next class of graduate students by the time they hung up. Fama would soon begin work on the dissertation that became a clarion call for David Booth.

In Fama's early years teaching, when Booth was one of the fervent students who packed his lecture halls, the professor wasn't much older than his disciples, but he was in much better shape. He biked, golfed, and windsurfed. He woke up at five A.M. and worked out to Wagner operas. He even played on Booth's intramural basketball team. "God, we were awful," Fama says. But he also carried himself with such gravitas that his students couldn't fathom a reality in which they were only a few years away from being the same age as this demigod. "I thought Fama was an old man," Booth says. "He was thirty." Booth had good taste in heroes. Many years later, when he really was an old man, Fama would be awarded the Nobel Prize in economics.

In his lectures Fama introduced a whole new set of phrases and terms to the lexicon of economics. One of them was that funny notion known as "market efficiency." That market prices already reflected all the relevant information about a stock and that it would be a waste of time pretending otherwise—that no single investor is smarter than the marketplace of many investors—was such an intoxicating concept that Fama's students couldn't wait to read the photocopied data that he distributed in every class. Fama was writing the book that would later help earn him the Nobel, and his students were getting the first drafts. All of it was so exciting that an animated Fama could leave the classroom windows open in the middle of a Chicago winter and sweat through his shirt. Booth idolized the guy.

"The most competitive person I've watched on TV is Michael Jordan," Booth says. "But the most competitive person I've ever been around is Gene Fama."

When I asked him why, he peered over his glasses. He proceeded to look at me like I had nine eyes.

"The notion of market efficiency has to be one of the most unpopular ideas on Wall Street ever."

Fama was molding a generation of David Booths into believing the philosophy behind every broker was more or less baloney. The people who claimed to be able to beat the market on a consistent basis were full of it, and there was no reason to give them your money. It was a form of intellectual courage for a scholar like Fama to challenge trillions of dollars in conventional wisdom, and unsurprisingly this made him an existential threat to the world's most powerful financial institutions. What he was saying, before anyone had the words to put it like this, was that it was silly to believe in the hot hand.

This was heresy. The whole point of the financial industry was to assume that you could beat the market. The skyscrapers on Wall Street were built on the promises of gifted stock pickers. While firms had to pledge in fine print that past performance was no guarantee of future results, you could almost feel them crossing their fingers behind their backs as they behaved otherwise. And it worked. The richest part of the richest country in the history of mankind might not exist if not for mankind's persistent belief in the hot hand.

Booth was bred to believe the opposite: that markets were efficient. It would have been hard for him to come out of this environment and *not* think the hot hand was an illusion. But the University of Chicago was like basketball for Booth. He loved everything about it. In terms of a career, it did not love him back. It didn't take long for him to notice that he did not have the personality to be a professor. Despite his enthusiasm for Fama, he nearly left school after his first Thanksgiving, when he felt there were turkeys having a better time than him. The following Christmas, he went to a relative's house in the sticks for the holiday. He took a quick break from his self-loathing to look around the dinner table at his family members. They were red from sunburn. They were missing an alarming number of teeth. They were in a home without indoor plumbing. But unlike him, they were happy. *What's wrong here?* he thought. He studied the dynamics of his family like a stock chart and reached a conclusion that none of the Nobel laureates he knew would have argued with.

"I realized they had life figured out," he said. "I was the one who needed to figure out what life was all about. That was the beginning of the end for me."

David Booth decided in that moment that he was done with the University of Chicago. He dropped out. There was no point of hanging around for a piece of sheepskin that proclaimed him to be a Ph.D. It would become clear to him many years later, when he'd made an unfathomable amount of money, so much that he could donate a small portion and have the University of Chicago's business school named after him, that he'd already been taught everything he needed to learn. Now it was time to apply it. Booth challenged himself to do something with the education he'd been given.

"One of the first things they teach you in business school is comparative advantage," he once said. "My comparative advantage was not in thinking up the next great idea. My comparative advantage was implementing the next great idea."

The next great idea waiting to be implemented was the belief that the hot hand doesn't exist in every industry. David Booth had the temerity to believe that betting against the hot hand could be a sensible business strategy. As it turned out, he wasn't the only one.

3.

Nick Hagen lives on a sugar beet farm on the border of Minnesota and North Dakota. He has a sugar beet tattooed on his tricep and a chromatic phrase from a Bach violin sonata tattooed on his bicep.

There was once a time when his diet consisted exclusively of spinach, peanut butter, and eggs, making him a wheat and sugar farmer who didn't eat wheat or sugar. All of which is one way of saying that Nick Hagen is not your average farmer.

But farming is the only business his family has ever known. The Hagens have been in the coldest part of the mainland United States since the presumably freezing day in 1876 that Nick's great-great-grandfather Bernt came from Norway and settled on the banks of the Red River on a patch of land straddling Grand Forks, North Dakota, and East Grand Forks, Minnesota. He built a small log cabin that later burned down, at which point he built a slightly bigger log cabin. It was as if he knew that his descendants would be sticking around for a while.

Nick was raised on the patch of land handpicked by Bernt. A fifth-generation farmer, he lives across the street from his father, who once lived across the street from *his* father. As a child, Nick understood that his family's sugar beet farm would always be there for him if he wanted it. He also understood that most children and grandchildren and great-grandchildren and great-great-grandchildren of farmers didn't have such a choice.

Nick planned to be a musician instead. He applied for the one spot in his class at Juilliard reserved for trombonists, and he practiced for seven hours a day before his tryout for the prestigious conservatory. He put so much stress on his body that he developed tendinitis. On the morning of his audition, he couldn't play without propping his arm on a desk in the hotel room. "I hope not to be able to say this for the rest of my life," he says, "but I have never been more sure and prepared and confident about anything in my life." He nailed the audition, moved from East Grand Forks to New York City, and barely left the blocks around Juilliard's campus. He coped with his homesickness by playing his trombone until the practice halls closed at midnight. His future wife described him as "a country bumpkin to the extreme, and an 80-year-old trapped in a young body." Nick was a farmer disguised as a musician.

He briefly went back to his roots on the farm after graduation, when he realized that he didn't want to do what he'd been training to do. Instead of becoming an orchestral trombonist, he became one of thousands of temporary workers who descend on the Red River Valley every fall to pluck thirty million pounds of sugar beets over the course of two hectic weeks. It was blissful. But not entirely convinced that farming was right for him, he moved back to New York after the beet harvest.

That's when he ran into his old classmate Molly Yeh. She had studied percussion at Juilliard before deciding she didn't want to do what she'd been trained to do, either. She was making her name as a food blogger and would become the star of her own Food Network television show. Nick and Molly knew each other casually from school, casually meaning they'd played Mahler's Fifth at Carnegie Hall, and she asked him about the visible tattoo on his tricep. He told her that it was a sugar beet. She asked him why he had a tattoo of a sugar beet. He recounted the long history of the Hagen farm. They started dating. Soon they decided they were ready to leave New York together.

The place they chose to spend the rest of their lives together was the Hagen farm. Nick's grandmother was getting too old to live on her own. Nick's father was getting ready to retire. It was now or never if he wanted to be a farmer. They chose now. Nick and Molly left the land of bagels for the part of the country that produced the hard red spring wheat in those bagels. His grandmother moved out of the house that his grandfather had built. Nick and Molly moved in. There were two Buicks in the garage and grain bins outside the front door. They got married on the farm in December when the temperature fell below zero degrees. They had their chickens and each other. It would have been totally idyllic if their timing hadn't been so uniquely terrible.

This was the beginning of Nick Hagen's farming education. The biggest takeaway of this schooling was one that he wouldn't and couldn't allow himself to forget. It was that he should never bring himself to believe in the hot hand.

Nick had come back to the farm at the moment when the stock charts of wheat and sugar beets had stopped making sense. The price of a commodity was supposed to go up and down, but the prices were only going up. It was bizarre. They had stumbled into such a boom market that even their cat, Sven, could've been a successful farmer. "All you had to do was look at the field and you made an insane profit," Nick says. He knew from growing up on the farm that prices didn't climb forever. What goes up and up and up must come down. They crashed right after he came back. He was too late for the boom and right in time for the bust. The dumb money that flooded the market washed away, and the people who bought freshly painted tractors and shiny pickups were the first to bite the dust. It would have been easy for Nick to think he'd made a huge mistake when sugar beet farmers sustained losses in his first two years back on the farm. That wasn't how he chose to see it. "I came into farming at just the right time," he says, "when the prospect for profitability was at its lowest." He didn't have a bull market to pad his bank account, but at least the biases that came with it would not infect his brain. He learned to stay the course, to save and not splurge, to play the long game.

That was the first lesson in the farming education of Nick Hagen.

Nick's transition from music to farming had been as much of a shock as moving from the boonies to the city. He'd always been obsessed with plan A because such obsession had been necessary in music. If he'd allowed himself to entertain the possibility of plan B, then he would be giving himself permission to fail. He thought of plan

B as something for people who couldn't master plan A. But when he came back to the farm, Nick found that his father was obsessed with plan B. He thought more about plan B than plan A. This flummoxed Nick. It took him a long time to accept that his life in farming would be the opposite of the life that he'd envisioned in music. Musicians had to be optimists. Farmers had to be realists.

Nick thought of himself as a trombone player, and being a trombone player meant striving to be the *best* trombone player. He couldn't think this way now that he was a farmer. "You can't be the best anything in farming," he says. "You're an okay mechanic, you're an okay agriculturalist, you're an okay businessman. I had to come to terms with the fact that being adequate was the objective." Nick had to be okay with being okay. It wasn't enough for him to accept failure. The goal was to embrace failure. "The more I think negatively, the more confident I am," he tells me one blistering summer day as we drive toward his wheat fields. "You feel settled and prepared and ready. Nothing is going to surprise you because you've already thought of all the terrible things that can happen."

That was the second lesson in Nick Hagen's farming education.

"I prepare for the worst," he says, "and hope for anything remotely better."

He did this by mitigating risk as much as possible. It wasn't much, and that only made it more important. "You can mitigate risk," Nick confesses, "by basically having no life." Nick has a wonderful life, but not when the sugar beets go into the ground in April and May, or when the wheat has to be picked in late August and September, or when the sugar beets are harvested from October 1 for however long it takes until they are out of the ground. "If my ground is fit to plant, no farmer who's using both sides of his brain is going to ever say, 'Well, maybe we'll just go away for the weekend,'" Nick says. "Here's the thing about agriculture. You're dealing with a live environment that's evolving and changing twenty-four hours a day."

After parking his pickup truck near the wheat fields, Nick spends the next hour fiddling with his combine. It isn't enough for him to basically have no life. He also mitigates risk with a maniacal compulsion about preventive maintenance. Before he plows hundreds of acres, he greases the stiff gears of his combine, and he proceeds to inspect the enormous piece of machinery like a Westminster Kennel Club judge examining a Pekingese. He'll work six hours in the fields this day. He'll put in a few more hours in the workshop behind his house as his chickens scamper in the yard. People ask him what he does in the winter. This is what he does. He tinkers. He picks apart every piece of equipment. He brings his combine to the shop for two weeks of nips and tucks. He even hires a professional to put it through the wringer of a hundred-point checklist of his own.

While this may sound like overkill, it's nothing compared to Nick's preparation for beet harvest. He's closer to paranoid than cautious when it comes to his precious sugar beets. Once each sugar beet gets beheaded, it's ripped out of the ground, dumped into the back of the harvester, and chauffeured to the nearest factory, where it's washed, sliced, and processed for the 17 percent of its body that is sugar. (A sugar beet exists to have the sugar beaten out of it.) By the time Nick and his crew are done, the fields are barren and the mountains of beets are small Alps. There is only one certainty in his deeply uncertain industry. The beets in Nick's backyard must be out of the ground by the time the frost appears and the deep freeze arrives.

We climb aboard his combine. It's sunny and hot with a cool breeze—a beautiful morning for wheat harvest. We are surrounded by golden fields as far as our eyes can see. It seems like there is no such thing as a horizon. Nick ignites the engine and starts driving. He is oddly relaxed for one of his busiest times of the year. He'd taken me to his favorite pizza spot the night before, and he is going to see a Prince cover band later that evening. Wheat harvest is hard work, but beet harvest is all-consuming. Molly likes to call beet harvest the "lunatic monster sister queen" of wheat harvest. And things can get a little nutty when the economy of an entire region depends on a lunatic monster sister queen.

This one stretch of the upper Midwest hugging the border of Minnesota and North Dakota is responsible for 10 percent of the country's sugar because the conditions here are ideal for precisely one specialty crop. Nick lives smack in the middle of Shangri-La for sugar beets. Nick's specialty crop insulates him from huge swings in the global commodity market. He invokes the wisdom of Warren Buffett to describe his good fortune: "I'm just really lucky that I won the sperm lottery that our farm happens to be two miles from the sugar factory and one mile from town in some of the most productive land in the world."

But to be comfortable in farming is to be vulnerable. Yes, there are fewer safer crops in farming than sugar beets. It's still farming. Nick has all the advantages that any sugar beet farmer could desire, and they've only made him more conscious of his disadvantages.

That led him to the third lesson of his farming education.

"Farming is defense," he says. "The most important variables are out of my control. I can get a good night's sleep, eat a hearty breakfast, plan my day to the minute, only to step outside and watch my crop get shredded by a hailstorm, or wilt under drought, or drown in a flood."

It had been that way for as long as Hagens have been farmers. When his first farm was decimated by pests, Bernt Hagen packed his bags, stuffed his life's savings in his pocket, and traveled the country looking for another spot for his next farm. The homestead that he chose was in almost the exact spot where his great-great-grandson now rides his combine. Nick learned to fear the weather long before it affected his daily livelihood. He was in fourth grade when a catastrophic flood ruined Grand Forks and its neighbors. Land became water. Trees became shrubs. Houses became docks. The rains came in April and closed school for the rest of the year. A once-in-a-lifetime disaster was enough to convince Nick of the unpredictable nature of farming. He didn't need another reminder. He got one anyway. In one of his first seasons back on the farm, biblical rain swamped Nick's field of sugar beets. It poured a few inches every few days, and Nick could see the storms from miles away. "Your heart would sink when you saw the sky turn dark," he says. He was in the muddy fields all hours of the day and night, and it still turned out to be the worst beet harvest that any of the living Hagens had ever seen. He could sympathize with his great-great-grandfather: Bernt Hagen also lost his crops one year during a violent storm that destroyed everything he owned.

Only a fool could delude himself into thinking that he has any modicum of control in a business where the only thing that really matters is as random as the weather. The weather is the single most influential determinant of a good year in farming. The whole business of farming sugar beets is like Stephen Curry attempting to play basketball on a court that can easily shrink or expand or morph into a rhombus. "The playing field is always shifting," Nick says. Farming is defense. Basketball is offense. "It is everything that farming is not and never can be," he says.

Nick is not in an environment that allows for the hot hand. In fact his environment actively punishes belief in the hot hand.

There is no point in believing that streaks of success have anything to do with his talent or circumstance. If he did behave that way, there would be a severe penalty. There is a very good chance that he would go broke. Nick can't have one good night in Madison Square Garden and change his business strategy. If he decides to scrap his sugar beets and bet the farm on soybeans, he would need to buy more planters, combines, and trucks; he would have to hope that soybeans remained profitable and that sugar beets had a few lousy years in a row; and he would have to pray the weather gods were on his side. The costs would almost certainly outweigh the benefits. The best-case scenario would be that it would take years of profit for the operation to pay for itself. "And by that time, soybeans may be a waste of time," he notes. The farming trends could have shifted back to sugar beets in the years that Nick pivoted to soybeans. Not even the latest advances in farming technology powered by sophisticated data science would be enough to help Nick. He could study every acre of his farm, notice a poor-yielding patch of land, and redistribute his seeds the following season. But his worst spot one year might be his best spot the next year. "That happens," he says. "Every year! It's always changing." His margin for error is about as thick as chaff. "That's the other thing about farming," Nick says, because one thing about farming is that there's always one more thing about farming. "Every season is unprecedented."

The wheat combine hums below us. Nick twists his body over the wheel and looks at his golden fields. There are hours to go and hundreds more acres to plow. He has plenty of time to reflect on the ultimate lesson of his farming education.

"I think you have to play principles over patterns," he says.

Nick has to constantly remind himself that one year's beet harvest has essentially no effect on the next year's beet harvest. Success *doesn't* beget success—at least not in farming. He couldn't acknowledge the hot hand if it slapped him in the face.

This abject lack of control is one reason that sugar beet harvest can be maddening. Nick drowns himself with nutritious lunches during harvest and still loses ten pounds over the course of two weeks. His daily life gets turned upside down, and then does a handstand, and then vaults into a backflip. At the end of the campaign, when the last crops are pulled, Nick is totally fried. "You're ready to never see a beet again," he says. In those stressful moments, when he feels at his most powerless, he occasionally wonders whether all of this is really worth it. But then he looks around and pauses to admire a night sky bright with stars and truck lights or the beauty of a sunrise on the Minnesota–North Dakota border. He stops and smells the beets. Nick comes to the same conclusion as his father, his grandfather, his great-grandfather, and his great-great-grandfather. Of course it's worth it.

And that is the last thing about farming.

"The stories of Grandma and Grandpa and Great-Grandma and Great-Grandpa are still part of your daily conversations," he says. "The lessons never leave."

Play defense. Remember the long game. Control what you can control. Prepare for the worst and hope for anything remotely better. Always stick to principles over patterns.

Nick Hagen isn't merely repeating the lessons of his ancestors. He is speaking the language of David Booth.

4.

David Booth's first office was his brownstone apartment in Brooklyn. The future financial heavyweight who would go on to manage more than $500 billion, give or take a few billion dollars on any given day, removed the sauna in his spare bedroom to install computer terminals and trading machines, and his phone company suspected he was a sports bookie when he requested that ten lines be installed as soon as possible. What the phone company couldn't have guessed was that Booth was about to revolutionize investing.

The cofounder of this company that operated from Booth's home was one of the few people he knew who shared his ideas about how markets really worked. Rex Sinquefield was raised in a Catholic orphanage and became enthralled with the stock market when he was supposed to be devoting himself to religious studies at seminary. The fact that he was more intrigued by money than Catholicism may have been one of the reasons that he never became a priest. Instead he enrolled at the University of Chicago's graduate school for business and became indoctrinated in the high church of market efficiency. Sinquefield had come to Chicago excited to leave with the secrets of beating the market. He left knowing that the secret was that there was no secret.

In retrospect it seems fated that Booth and Sinquefield would have fallen under the spell of Gene Fama. But it wasn't. A great majority of American investors thought he was profoundly wrong. "I'd compare stock-pickers to astrologers," Fama liked to say. "But I don't want to bad-mouth the astrologers." I once asked Fama if he ever looked around at all the people who put their money in the hands of these professional stock pickers and thought, *Are you people absolutely out of your minds?* "They're not absolutely out of their minds," he said. "They just don't understand statistics."

Fama got Booth his first job in finance after he dropped out of school. He'd become friendly with a man named John "Mac" Mc-Quown, one of the other people who believed Fama was right and the banks were wrong, which was a bold stance considering McQuown worked for a bank. McQuown was in the process of helping pioneer the investment vehicle now known as the index fund, a safe place for investors to dump their money and watch it multiply over time, and Booth was recruited to help him. Sinquefield was busy pursuing a similar project for a competing bank. Their index funds weren't complicated or purposefully obtuse. They simply tracked the Standard & Poor's 500 and more or less mirrored the results. But what made them such radical departures from the existing norms was the way they contradicted the advice of the entire financial industry. The essential folly of the banks was that their smartest and most valued people were doing a whole lot of meticulous work that amounted to the same result as doing nothing. In fact doing nothing would have been more profitable. When they subtracted the fees they collected for doing something, the banks were providing their clients a *worse* product. A low-cost service that reflected the gains of the market and charged less than active managers seemed like the smarter play to those students who had studied under Fama. The machinations of the financial system were so complex they were content outsourcing the work to the market itself. But their colleagues still believed they possessed the ingenuity to outperform an index fund despite all the evidence to the contrary.

Once they were tired of being the resident heretics inside banks, Booth and Sinquefield set off to start their own company using their deeply unsexy approach to investing. They pitched their potential clients on diversifying their portfolios and holding the market by giving themselves more exposure to smaller companies. They didn't employ research analysts or celebrity investors, nor did they have any of the other trappings of a Wall Street behemoth in the global headquarters of Dimensional Fund Advisers, which happened to be Booth's apartment. They simply created an index fund consisting of the smallest companies on the market. But the truth is there was nothing simple about it. They combined a little bit of judgment with the knowledge that any more confidence in their own judgment was dangerous. "In every other industry, if you have smarter people who work harder, they'll do a better job," Booth says. But not in his bizarre world of investing. Those smart people will do a better job if they stop working so hard. It's not productive to outguess the market. In fact it's counterproductive. That's what Booth thought, anyway, and his track record has forced people to come around to his way of thinking. "The first time people hear it, they think that can't possibly be true," he once mused. "Now we're almost mainstream. Heaven forbid."

Booth also believed in principles over patterns. He believed that streaks were aberrations. He believed that betting on them was riskier than bowing to probability. He believed that passive investing was far superior to active investing. And he believed that what he believed was not all that innovative. "To my mind," he says, "it's Economics 101." This required Booth to check his ego. Then again he was a Kansan. He didn't have an ego to check. "If there is a genius out there, it's not me," he admits. "That's all I know."

But come to think of it, he did know one genius. The first person he called when he started Dimensional was Fama. He asked his former professor to be on the company's board, figuring it would be smart to have Fama on the payroll if they were going to put his lessons to the ultimate test. Fama always thought Booth would do well after graduate school, but they hadn't been in contact since he dropped out. Booth hatched the idea for his firm in 1980. Dimensional was incorporated in the spring of 1981. In the fall of 1981, Fama happened to be advising an interesting graduate thesis. This thesis identified an unlikely investment opportunity: the most piddling stocks on the market. The theory was that smaller companies outperform larger companies over the long term. Dimensional turned this theory into a strategy. Booth's startup was no longer advising clients to diversify their portfolios. Now it was telling them that diversification would lead to higher expected returns.

There was another study that was published after the founding of Dimensional that would vindicate its approach. The authors of this famous study found that throwing darts at the stock pages of the *Wall Street Journal* would yield as much profit as hiring a portfolio manager. Booth liked to embellish the research in a way that only underscored the point: What if the people throwing the darts were orangutans instead?

"One orangutan in a thousand will beat the market every year for ten years in a row," he said.

That orangutan would be hailed as a prophet. He would be a fixture on CNBC. The financial press would publish fawning profiles of him. There would be no doubt that this particular orangutan had the hot hand. Only when you replace "human being" with "orangutan" does that set of events begin to sound ludicrous.

"Most MBAs think they're that one special orangutan," Booth said.

And most investors want to believe them. That was the whole point of Gilovich, Vallone, and Tversky debunking the hot hand in basketball. The "powerful and widely shared cognitive illusion" they uncovered was powerful and widely shared because it was not limited to basketball. They weren't saying those orangutans didn't exist. Of course they exist! One paper looked at more than five thousand mutual funds between 1962 and 2008 and found about 4 percent of those funds enjoyed winning streaks that couldn't be explained by sheer chance. There *was* skill. But there were also orangutans. "That's really what's difficult for people to accept," Booth says. "What am I supposed to do? Pick the guy with terrible numbers? How am I supposed to pick a manager if I don't look at historical returns? The answer is that you shouldn't be picking managers."

Booth has been in this argument enough times in his life that by now he's developed a skill of his own: mind reading. He knows what you're thinking. "Okay, I'll give you Warren Buffett," he says. "Name me another one."

Not even Warren Buffett would want any part of that bet. Buffett once made a $1 million wager with a hedge fund manager named Ted Seides, and the most successful stock picker in the history of money took Booth's position. The showdown was hatched in earnest at Berkshire Hathaway's annual meeting in 2006. Buffett took the stage in Omaha with his trusted partner Charlie Munger. Given the shrieks they elicited, they could have been confused with John Lennon and Paul McCartney. After outlining their investment strategy to the adoring crowd, they held court for hours as Buffett fans shouted questions at him. But it was the very last question of the afternoon that yielded the most revealing answer from the man they had come to see. The question itself was rather dull. It was one that he'd been asked thousands of times: What was Warren Buffett's advice for all those people who wanted to be Warren Buffett?

Finance was a funny business, he said, because the very thing you're paid to do is actually the thing you shouldn't do. When you have a baby, you want an obstetrician by your side. When you clog the toilet, you hire a plumber. When you get locked out in the middle of the night, you call the locksmith. "Most professions have value added to them above what the laymen can accomplish themselves," Buffett said. "The investment profession does not do that."

He offered to back up his bravado in the most Warren Buffett way possible: by opening his wallet. Buffett predicted that active management would underperform passive management over the long haul. This was an unexpected stance for a stock picker. But his theory, which was Booth's theory and the theory of an increasing number of index fund investors, was that the fees that managers charged to justify their work made it harder for them to spin a profit for their investors. Buffett thought investment fees were a scourge of the industry, regarding them with all the esteem of bedbugs, and he put his money where his mouth was. He plunked down his own cash in a public wager that would take place over ten years. He took an unmanaged index fund and waited for hedge fund managers and stock pickers to line up for their chance to prove him wrong.

"What followed was the sound of silence," Buffett later wrote. "These managers urged *others* to bet billions on their abilities. Why should they fear putting a little of their own money on the line?"

Seides was the only person chivalrous enough to defend the honor of his industry. He chose a carefully selected portfolio of hedge funds that he projected to outperform the securities market over a period of ten years. Buffett was so confident that he even let Seides tweak his approach to adjust for the latest shifts in the market. It was almost as if Buffett were encouraging his opponent to chase the hot investors and dump the cold ones.

Their bet started on January 1, 2008. It was supposed to end on December 31, 2017. Buffett declared victory with one year remaining. It was such a rout that Seides could no longer catch him. The $1 million invested in the fanciest hedge funds selected by an expert would have returned $220,000. The same amount invested in a boring old index fund that Buffett ignored for a decade would have returned $854,000.

Buffett reported the results of their bet in one of his famously enjoyable shareholder letters. This was the written version of a victory lap. "I've often been asked for investment advice, and in the process of answering I've learned a good deal about human behavior," he wrote. "My regular recommendation has been a low-cost S&P 500 index fund." Buffett took his advice to heart. His will lays out explicit instructions for how to invest his fortune when he dies: put 90 percent of his cash in one of those low-cost S&P 500 index funds. He once summed up this philosophy in a way that both David Booth and Nick Hagen would appreciate. "Ignore the chatter, keep your costs minimal, and invest in stocks as you would in a farm," Buffett wrote.

Warren Buffett wasn't the only investor to win a bet. David Booth did, too. Never before had anyone become so wealthy on the strength of a single idea.

Booth moved his offices to Los Angeles when Dimensional became too big for Brooklyn. The company's investing strategy evolved over the years, and by the time it moved again to Austin, Dimensional was one of the richest asset management firms known to mankind. But that's not how Booth chose to think about it. To him it was more like a chain of sandwich shops. "We would only be about 18 Jimmy John stores," he bragged.

It should go without saying by now that, for the most part, David Booth doesn't carry himself like a financial celebrity, nor was he treated as one for most of his career. He works from nine A.M. to five P.M. He goes home when he gets tired. He tried sitting on an exercise ball in the office but it made his lower back hurt, and he went back to a plain old swivel chair. He doesn't own a sports team. He's not building spaceships. He's as close to anonymous as a billionaire can get, which is oddly fitting, given that he amassed his fortune by constantly reminding himself to ignore his ego. The first time that anyone outside high finance heard the name of this delightfully square investor was when he donated $300 million to the University of Chicago to have the business school named after him. In the *Wall Street Journal* article reporting the news of the gift, Booth is described as "largely unknown outside the rarefied world of academic research." Even the most prestigious business publication had to explain who he was. But it was worth burning his anonymity to give back to the place that had trained him to think this way in the first place.

"If you look at what's the fundamental question in investing, after all these years of research by academics, in some senses it really is: Are there hot hands in stock picking?" he says.

David Booth was one of the first people to say no.

"I say the best working assumption is no," he concludes. "There could be. But we just don't know how to identify them before the fact. We don't know how to identify which orangutan will be the winning dart thrower."

5.

It had already been an excellent day for Sotheby's auction house by the time David Redden opened the bidding for the final lot on a chilly afternoon in December 2010.

The auctioneer had sold an early copy of Thomas Jefferson's *Notes on the State of Virginia,* Bob Dylan's original handwritten lyrics for "The Times They Are A-Changin'," and the collected plays of our old pal William Shakespeare. There were three items left on the docket after lunch. The first was Robert Kennedy's copy of the Emancipation Proclamation that was so clearly a treasure that not even Sotheby's bothered hyping it. The second was a flag of George Custer's army that Sotheby's called "the most significant and symbolic artifact recovered from the Little Bighorn Battlefield." But the third and final lot of the afternoon was, according to Sotheby's officials, "a document that transcends sports, being the genesis of a creation of American culture that has become as influential as jazz and as pervasive as Hollywood." For sale were James Naismith's original rules of basketball.

Redden was used to holding the gavel for such momentous auctions. He liked to say that his specialty was everything from spaceships to dinosaurs. He wasn't exaggerating: Redden *had* sold a lunar spaceship and *Tyrannosaurus rex* fossils. He'd also sold Mozart symphonies and Einstein formulas, Fabergé eggs and Andy Warhol cookie jars, the most valuable stamp, and the most valuable coin at that time. He'd even sold what Sotheby's called the most important document in the world: the last privately owned copy of the Magna Carta.

The winner of that Magna Carta was a man of extravagant means named David Rubenstein. He used a fraction of his fortune on this sheet of parchment and promised afterward that he would've paid more than the $19 million it cost if that had been necessary. ("I don't think you can put a price on freedom," he said.) Rubenstein collected historical artifacts like teenage boys collected baseball cards and comic books. He once described his buying strategy as: "I buy everything." Everything meant that in addition to the Magna Carta, he also owned copies of the Declaration of Independence and the Constitution.

He'd already purchased one of Abraham Lincoln's autographed souvenir copies of the Emancipation Proclamation a few years earlier and loaned it to the White House, where it hung above Martin Luther King Jr.'s bust on a wall in Barack Obama's office, and now he was in a private room inside the Sotheby's building to bid on another one. But if he'd gotten his way, he wouldn't have been there. Rubenstein had contacted Ethel Kennedy directly to make a private offer, but Robert Kennedy's widow decided to proceed with the auction as planned. She told Rubenstein she figured it might fetch more. She was right. Redden whipped the potential buyers into a bidding frenzy that pushed the price of the Emancipation Proclamation all the way to $3.8 million. "As if Abraham Lincoln was going to show up and sign it again," Rubenstein said. "It went way beyond what I thought it was worth." Rubenstein dropped out. "I was disconsolate," he said.

His liaison from Sotheby's tried to cheer him up. She knew that Rubenstein liked basketball. She also knew that the next auction was for Naismith's rules of basketball. "Why don't you buy it?" she said.

Rubenstein was vaguely familiar with the history of the sport's invention and figured the rules would be a nice consolation prize to the Emancipation Proclamation. They didn't abolish slavery in the United States, but they *were* intriguing in their own right. "One day there wasn't basketball," a Sotheby's official said. "The next day there was." Rubenstein agreed to stick around.

"The third and final sale this afternoon," Redden said to his rapt audience, which unexpectedly included Rubenstein. "The founding rules of basketball!"

Redden opened the bidding at $1.3 million. The price soared to $1.4 million, $1.5 million, $1.6 million, and $1.7 million. The auctioneer could barely contain his excitement. There was a certain thrill in not knowing the identities of these people committing tremendous amounts of money in rapid succession. It could be someone like Rubenstein—or it could be someone unknown in an office building thousands of miles away.

"When I want to bid," one of those bidders asked his phone attendant, "how do you want me to handle it?"

"We're at 1.5, 1.6, 1.7," the phone attendant said. "Would you like to say 1.8?"

"Yes, I would, please."

Rubenstein suddenly had some competition.

"1.9," the phone attendant said. "Would you say 2, sir?"

"Yes."

"We're at $2 million with you."

"Well, don't let anybody else bid," the anonymous bidder said.

"2.1," she said. "Would you say 2.2?"

"Yes," he said.

"2.3," she said. "Would you say 2.4?"

"Yes."

"2.5. Would you say 2.6?"

"Yes."

The original rules of basketball were now officially worth more than Custer's flag. The buzz in Sotheby's built as these two bidders with deep pockets went back and forth. The higher the price, the fewer people it could be. There were only so many billionaires who could afford to spend millions of dollars on a sheet of paper.

Rubenstein was taking his sweet time with his bids. But this other guy, whoever he was, wherever he was, whatever his purpose was, said yes to the next $100,000 increment with all the deliberation of whether he wanted fries or salad. Rubenstein had been around enough auctions to understand this meant he was most likely doomed. "If someone is bidding against you and rapidly increasing your price," he said, "they're going to win." But not quite yet. Rubenstein upped the price to $2.7 million.

"Would you say 2.8?"

"Yes," said the man on the other end of the phone line.

"They went to 2.9," the attendant said. "Would you say $3 million, sir?"

"Yes."

Rubenstein was intensely curious about who else might be willing to spend the average price of an Emancipation Proclamation on the rules of basketball. "Sometimes, if I'm buying the Magna Carta or the Constitution, I know the people who care about this kind of stuff," he said. "But it's very hard to know for sure." It was even harder in this case because Rubenstein wasn't a usual buyer of sports memorabilia. He had a hunch that he might be competing against Nike founder Phil Knight, but that was the only guess he could muster. He was too busy trying to keep up with this mysterious rival pushing the price of something he hadn't considered buying until a few minutes earlier beyond his limits. "Generally, when you go into an auction, you think this is what I'll pay, and you wind up paying twice that," he said. That had been a few million dollars ago. And the guy on the other phone line was getting impatient. Rubenstein called $3.1 million.

"Would you say 3.2?"

"Yes."

"3.3," the phone attendant said. "Would you say 3.4, sir?"

"Yes," he said. "But tell them to speed it up if you wouldn't mind."

"They went to 3.5. Do you want to say 3.6?"

"Well, let me think about it," he replied. "I don't know if I want to do it or not."

The man on the other end of the phone thought about it for long enough to make it seem like he was actually thinking about it.

"Okay, 3.6," he said. "Is he going to drop out or bid? I don't know what's so complicated. What's he doing?"

"They went to 3.7," she said.

"Okay, 3.8," he said.

That was it for Rubenstein. It was time to let this mysterious guy have what he so badly wanted. "It was apparent that whoever was on the other side was not going to lose," he said. Rubenstein was going home empty-handed. He couldn't put a price on freedom. But he could put one on the rules of basketball. It was $3.8 million.

"At three million eight hundred thousand, are we all through?" Redden said. "Definitely through? Three million eight hundred thousand. On my left—at three million eight hundred thousand dollars!"

"It's yours, sir," the phone attendant said.

"Okay, great," said the anonymous bidder.

Rubenstein could no longer stand the suspense. As soon as the auction was over, he had to know more. Who was this mysterious bidder?

He felt better as soon as the anonymous winner revealed his identity. The competition was not who Rubenstein was imagining. He was richer.

Rubenstein already had a relationship with this other very rich man on the phone. They were on the board of trustees of the University of Chicago together, as very rich men tend to be, and Rubenstein learned that his rival wanted to buy the original rules of basketball and donate them to his alma mater: the University of Kansas. He'd been raised in a house so close to campus that his crib might as well have been in a dormitory. The exact address of his childhood home was 1931 Naismith Drive.

The billionaire who bought the original rules of basketball was David Booth.

Five

WHEEL OF FORTUNE

"It's my life."

1.

"That's me!" says the famous Iraqi sculptor.

Alaa Al-Saffar is in the artist's studio connected to his small home on a sunny California day when he points to a faded photograph on his table. It is a portrait from his old life working for a murderous dictator. The man in the photo wearing blue jeans and with hair below his shoulders is the only person in a room of Iraqi officials not dressed in a suit or a military uniform. He looks hilariously lost. It is as if he'd been on his way to a gallery opening and stumbled into a cabinet meeting. The effect of Al-Saffar's presence is so disorienting that if you were to glance at this photo you would notice him before you realized the man standing beside him is Saddam Hussein.

It was neither by accident nor by choice that Alaa Al-Saffar found himself working for one of the most brutal leaders of the twentieth century. A few times every year, Al-Saffar received an envelope stuffed with cash. He didn't need to read the note inside to know that it was an invitation from the president of Iraq. "Saddam was an easy person to get on with," Al-Saffar says. "At least with us artists." But his personal and professional opinions about this man nicknamed the Butcher of Baghdad didn't really matter. He'd been summoned enough to understand that honoring the whims of Hussein was a minor inconvenience that Iraqi artists had to deal with every so often. It was like a teeth cleaning.

He tolerated Hussein as a patron only because it allowed him to pursue the life of an artist. The son of a painter, Al-Saffar studied art in Baghdad and left his native country to pursue a master's degree in Switzerland, which afforded him the opportunity to visit Paris and Rome to study Rodin and Picasso, Chagall and Dalí, Michelangelo and Da Vinci. He returned to Baghdad in the 1980s and immediately distinguished himself by winning a series of prizes for his sculptures, paintings, and sketches. He would never have the stability of a steady job or a reliable paycheck, and there might be some months when he worked from the garage of his Baghdad home because he couldn't afford his studio. But he was fine with the choices he made. He accepted that his commitment to his art required sacrifices. "He worked nonstop," recalls Zinah Al-Saffar, the oldest of his children. "His dream was to have a big statue in the United States or have his paintings on display in one of the biggest galleries in the world."

But that's when he became one of the dictator's favorite artists. The attention of Iraq's president was a powerful amplifier that could make or break careers, and when Hussein invited artists to private competitions, they didn't think twice about taking his money. "It's not that you are forced," says Natiq Al-Alousi, another Iraqi artist. "It's up to you whether you want to participate or not. And everybody wants to participate. That's income. That's work. Everybody is looking for work." By the time his presence was requested in late 2002, the year before the United States invaded his homeland, Al-Saffar was used to taking a commission from someone who used chemical weapons against his own people while treating artists like heroes. He was especially proud of the last sculpture that he proposed to this man who was about to burrow into a spider hole. It was meant to be a gargantuan bronze piece to reflect his national pride. The sculpture would be his most ambitious project yet. On top of the whole thing, like the bride and groom figurines on a wedding cake, was an enormous palm tree. "Iraq is famous for the palm," Al-Saffar says. "Like California."

Hussein thought it was almost perfect. But the dictator had some notes. He suggested one thing that he felt was missing from the sculpture: a massive statue of Saddam Hussein atop the palm. Al-Saffar knew what to say: "I say okay," he said. He agreed that his sculpture could use a massive statue of Saddam Hussein standing on a palm tree.

But the war began. Saddam Hussein was captured and killed. Alaa Al-Saffar's sculpture would never come to exist beyond the scope of his imagination.

Instead he spent the days after Hussein's downfall chiseling stone in his garden and making art for which there was no market. Al-Saffar felt like he was on the verge of a breakthrough when he won a national commission for a memorial sculpture, but the funding for that project went dry before he could start his work, and then his life fell apart because of it. When his name was publicly mentioned in the news for winning the commission, he became a target of Al-Qaeda and the insurgent militant groups as they consolidated power in the aftermath of Hussein's reign. (As it turned out, Al-Qaeda wasn't full of art buffs.) It was suddenly dangerous to be Alaa Al-Saffar.

His daughter was getting ready to bring her children to school one day when she says she noticed an envelope slipped under the garage door of the family's home. This one wasn't stuffed with cash. Inside the envelope was a death threat for her father. "It was an order for him to stop being an artist," Zinah Al-Saffar says. "Not a request."

There would be no more painting and sculpting in his garden. If he kept working, he would be killed. "They called me an infidel and told me to stop," he said. "So I stopped."

Zinah soon moved to Southern California with her family because her husband had worked for the U.S. Army during the war, and she had to watch from afar as her father struggled to adjust to this life without art. She could see how tenuous his situation was. Every time she called home, he tried to reassure her that he was safe. "But I knew," she says. "I was there when he got the letter. I'm the one who found it."

She urged him to leave. "You don't belong there!" she told him. He wouldn't listen. When she finally persuaded him to visit, he landed in Los Angeles with a temporary visa. His trip was only supposed to be a vacation until they found themselves staring down the worst part of vacation: the end. Zinah couldn't bear the idea of her endangered father returning home. "Dad, I don't think you should go back," she said. "Stay here."

He knew that she was right. In that moment he accepted that his old life was over. The famous Iraqi sculptor was in his sixties when he decided it was time for a new one.

He filed the paperwork to inform the proper authorities that he was in grave peril back home, that he feared persecution if he returned to his native land, and that he needed to be protected by the United States. And thus began the asylum process for the artist from Baghdad. Alaa Al-Saffar's new life was about to be seized by the hot hand.

2.

Justin Grimm wasn't sure why he was being called into his boss's office. It was another oppressive day in Frisco, Texas, the latest stop on the carousel of minor-league baseball teams Grimm had been riding since he'd been drafted by the Texas Rangers. He'd already played for the Hickory Crawdads and Myrtle Beach Pelicans. Now he was a

Frisco RoughRider. He expected to be in Double-A ball for a while after the Rangers had invited pretty much everyone with a realistic shot at playing for them to spring training that season and had not invited him. But he'd been pitching well lately, and the people who drafted him were paying attention, which was why his manager needed to see him on this particular Thursday afternoon.

The Texas Rangers needed a starting pitcher on Saturday night against the Houston Astros. They decided against every reasonable expectation, including Justin Grimm's, that the man for the job was Justin Grimm.

In front of a sellout crowd of nearly fifty thousand people, including a former president of the United States, he soon found himself strolling to the mound with wobbling knees as the television announcer said to the audience at home, "Who knows what's in the stomach of Justin Grimm?" As it turned out, not much. Grimm was so nervous that he hadn't eaten all day. He toed the rubber, took a deep breath, tricked himself into thinking his big-league debut was just like any other game, stared at his catcher for the sign, and fired a ninety-one-mile-per-hour fastball. His ability to throw a baseball with amazing velocity and pinpoint accuracy was one of the reasons the Rangers thought he was ready to leap from Double-A on two days' notice, but he would later admit that he had no earthly idea where the ball was going when it left his right hand. It zipped down the middle for a called strike. The first pitch of Justin Grimm's major-league career was perfect.

Grimm let the leadoff hitter get on base with a single, but he got the next hitter out, and he was beginning to settle down and forget that he was living his childhood dreams when Jed Lowrie walked to the plate.

For a professional athlete, Jed Lowrie was awfully peculiar. He might have been the best player on the worst team in baseball that season. The most striking physical attribute of this undersized six-

foot, 180-pound shortstop was his piercing blue eyes. He represented
the few dozen professional baseball players with a college degree.
And he wouldn't have been in the batter's box that day against Justin Grimm if he didn't have one.

Lowrie was a political science major at Stanford University the
first time an improbable major-league scout named Sig Mejdal came
to see him. Mejdal studied aeronautical engineering in college and
operations research and cognitive psychology in graduate school,
and he paid his tuition through a side gig as a blackjack dealer at
the local casino. He was working for Lockheed Martin and NASA
as a literal rocket scientist when he decided that he wanted to work
for a baseball team instead. His first job in baseball was in fantasy
baseball: Mejdal was the quantitative analyst for a lunatic owner in
a wildly competitive fantasy league. That invaluable experience led
to a position in the fledgling statistical department of the St. Louis
Cardinals—a real job with a real team. Mejdal built a model that
used a prospect's college statistics to project his future in the big
leagues, and his computer told him the best college baseball player
in the country was playing at Stanford.

But when he went to see that player in person, the most remarkable thing about Lowrie was how unremarkable he was. Jed Lowrie
resembled a Major League Baseball player as much as Sig Mejdal carried himself like a grizzled scout. Mejdal resisted the urge to break
his laptop for long enough to remember that he didn't care how Lowrie looked, only how he played, and he played the way his algorithm
suggested. Mejdal begged his bosses to draft Lowrie. When they
passed on him with their first pick, Lowrie was selected by the Boston Red Sox instead. It would soon become clear that the Cardinals
should have listened to Mejdal. Lowrie made it to the big leagues,
just as the algorithm said he would, and he was coming into his own
when Mejdal's boss was named the general manager of the Astros
on December 8, 2011. Less than a week later, he traded for Lowrie.

Less than a month later, he poached Mejdal. Finally they were on the same MLB team.

By the afternoon of June 16, 2012, when he was staring at Justin Grimm on the mound, Jed Lowrie was once again proving Sig Mejdal right. He was on pace to hit more homers in his first season with the Astros than in his previous four seasons combined. He'd already hit eight home runs that month after never having smashed more than nine in a full season. But that was mostly because it had never really been his goal to hit home runs. When he was in high school, Lowrie's field didn't have a fence. There was no incentive for him to hit the ball high and far in the air. The only way that he could guarantee himself a home run was by hitting it so hard that it kept rolling long enough for him to round the bases. That became his goal: hit the ball as hard as he could.

But the sport of baseball was on the brink of its own version of the three-point revolution that put a heavy premium on strikeouts for pitchers and home runs for batters. This one at-bat in this one meaningless game between the Rangers and Astros was an unexpected peek at the future. Grimm wanted a strikeout. Lowrie wanted a home run.

Grimm's parents were escorted to the best seats in the house, right behind home plate, in time to watch their son throw the first pitch to Lowrie and hear the umpire Bill Miller call the fastball a strike. There was nothing memorable about this pitch. It was one of the hundreds of thousands of pitches that make up a baseball season. Lowrie rearranged the dirt in the batter's box. Grimm looked at his catcher for the sign and conveniently ignored the fact that he was about to throw his next pitch as hard as he could almost directly at his family.

It was around this time when the showdown between Grimm and Lowrie subjected itself to the fickle biases of human judgment.

As soon as he took the mound, Grimm had the advantage over

Lowrie. The pitcher always has the advantage over the batter in a game where even the best hitters fail more than they succeed. But his advantage in this particular situation was even bigger. Grimm had never pitched in the big leagues, and there was no reliable scouting report for Lowrie to study. He wouldn't know how the pitches would look until they were hurtling at him. "You don't know what they *do*," he says. Grimm's inexperience meant that Lowrie was basically wearing blinders when he stepped to the plate.

The odds of this at-bat shifted further in Grimm's favor as he prepared to deliver the next pitch. The catcher Mike Napoli moved several inches away from Lowrie. By crouching on the outside corner, he was attempting to fool the umpire. Miller might be tricked into calling a ball a strike if the pitcher didn't make the catcher move. Grimm and Napoli were conspiring to make the strike zone slightly bigger. It worked.

Grimm put the ball where his catcher wanted: low and outside. Lowrie recognized that it was low and outside and figured that it would be called a ball. Miller called a strike. Lowrie couldn't believe it. He turned his whole body and registered his discontent with a vicious stare. The silent protest was Lowrie's equivalent of screaming in the umpire's face given how rarely he argued the merits of a called strike. "Only when it's a ball," he mutters to me years later.

By this point of his umpiring career, Bill Miller was impervious to Jed Lowrie's stink eye. Miller was basically a lifer behind the plate. He first started thinking about umpiring when he was still in middle school. He made some extra cash in high school by calling Little League games, quit his college baseball team to be a high school umpire, and went to umpiring school to work in the minor leagues as soon as he could. He'd been in the big leagues fifteen years, and he loved his job. He loved making a judgment after every pitch, and he especially loved coming to Texas, where the clubhouse attendants Hoggy and Cornbread made sure there were always heaps of jumbo shrimp, brisket, and apple crisp waiting for him as he took off his uniform. He was even used to the one part of umpiring that could make him not love his job: the complaints. To complain about balls and strikes was as much a part of the game as the buckets of sunflower seeds in every dugout. There were times when it seemed like the purpose of Miller's job was to call pitches and then get berated about how bad he was at his job. A good umpire had to acknowledge his mistakes, however, and Miller was objectively a good umpire.

But there was one more thing about Miller that made his mere presence behind the plate better for Grimm than Lowrie. He was a friend of pitchers. Miller called about four more strikes per game than the average umpire. In fact his strike zone was among the biggest in all of baseball. "High pitches, low pitches, inside pitches, outside pitches, pitches to left-handed hitters, pitches to right-handed hitters—Miller is almost always calling more strikes," read one description of his style on the baseball blog *Hardball Times*.

It wasn't possible to say with definitive proof whether any umpire was right or wrong when Miller made it to the big leagues. But then came a system called PITCHf/x that planted high-resolution cameras in every ballpark and tracked in precise detail everything you could have ever wanted to know about any given pitch: its speed, its trajectory, and, most important, its location. It was technology that could make people like Bill Miller obsolete. Human error had always been a part of baseball if for no other reason than it had to be. But now it didn't. There was nothing other than the inertia of a hidebound game that was stopping Miller from being replaced with a machine. In the meantime, Miller's bosses developed a method known as Zone Evaluation, which used PITCHf/x data as the unimpeachable standard to determine how often umpires like Miller were right and wrong. It was an ominous reminder that the robots would never be wrong. But until they came along, balls and strikes would still be called by humans. The upshot of PITCHf/x was that it was suddenly possible for anyone, not just Miller or his bosses but everyone with access to the internet, to see for themselves if a pitch was a ball or a strike as soon as it crossed the plate. In this particular instance, the umpire believed the blur of a fastball had grazed the outside corner of the plate, but really it had missed the plate by an infinitesimal amount that made all the difference.

It was called a strike. It should have been a ball. Jed Lowrie was right. Bill Miller was wrong. Justin Grimm was lucky.

Lowrie now had two strikes against him as Grimm looked at his catcher for the sign and saw one finger: the pinkie. A pinkie in the major leagues is the same as the pinkie in Double-A ball and every level of the game all the way down to Little League. In the universal language of baseball, the pinkie is an inside fastball. Grimm nodded in agreement with his catcher. They both felt the inside fastball was the pitch that would strike out Lowrie. The cameras in the stadium that night would record this fastball at ninety-three miles per hour as it whirred over the inside edge. The precise baseball terminology for this kind of pitch is "filthy."

If the umpire were a robot making decisions using robot logic according to a box over the plate that only a robot could see with robot eyes, he would have initiated a strike-three call after such a filthy pitch.

Bill Miller was not a robot. He was very much a human being. And the umpire with the biggest strike zone in the game called this borderline pitch a ball.

The one thing that's hard to appreciate about MLB umpires when drunk fans are screaming obscenities at them is that they are phenomenally good at their jobs. They manage to nail about 87 percent of their calls, and they almost never miss the obvious balls and strikes. Their success rate on those pitches is 99 percent. It's the toss-up calls within inches of the strike zone that give them fits. Even the most reliable umpires are right about those pitches only 60 percent of the time.

But that wasn't the only reason Miller called the pitch from Grimm to Lowrie a ball. A few years later, this pitch was one of millions over five MLB seasons analyzed by a team of economists intrigued by the behavior of professional decision-makers in high-stakes situations. What they wanted to know was how umpires acted immediately after calling two consecutive strikes. And what they really wanted to know was whether it had any effect on the next pitch.

In a perfect world, it wouldn't. Major League Baseball is not that perfect world. If the last two pitches were called strikes by the umpire, he was 2.1 percentage points less likely to call a strike. Miller's strike zone had shrunk for this pitch. The fact that Miller had already called two strikes and Grimm had thrown another near-strike actually conspired against him.

Grimm betrayed no emotions about being robbed of the first strikeout of his big-league career. He pursed his lips and stared at the catcher for his next sign. There it was again: the pinkie. He didn't need to nod this time. As soon as he saw that pinkie, he went back to his set and fired another inside fastball. The runner on first base took off for second base. The catcher threw a perfect strike of his own to catch him stealing for the second out. Grimm exhaled again. He was now one pitch away from the end of the inning.

Grimm went into his windup and uncorked a ninety-four-mile-per-hour fastball right over the heart of the plate. It turned out to be a huge mistake. Lowrie pounced on the second chance afforded to him by Miller not calling a third strike. The instincts that he'd been training since his days in high school on a field without a fence took over. He swung. The ball cracked off his bat and climbed for the abyss between right and center field. Grimm's mother covered her mouth in horror. But the ball was still going. Grimm's sister had shielded her eyes at the sight of contact. And still the ball kept going. She finally allowed herself to peek at the worst possible moment: when it sailed over the fence.

Jed Lowrie hadn't just hit the ball hard. He'd hit a home run.

3.

The one thing that Justin Grimm, Jed Lowrie, and Alaa Al-Saffar have in common is that each of their predicaments can be explained by taking a trip to a casino.

A casino is a good place to study decision-making because casinos are where people go to make bad decisions. This laboratory of willful stupidity has the controls that an experimental psychologist desires, but it also lures subjects who are real people betting their own real money, which is why every spin of the roulette wheel has all the makings of a rollicking study of human behavior waiting to be written. Rachel Croson and Jim Sundali decided to write it.

Croson was a behavioral economist who focused on financial decision-making. The kinds of mistakes that people make in their dealings with money were the cognitive errors that power the entire casino industry. Sundali knew that from firsthand experience. Before he was a scholar of managerial sciences, he was a stockbroker. When he went back to school for his Ph.D. he studied under an Israeli psychologist named Amnon Rapoport, who had been college roommates and inseparable friends with Amos Tversky. They had met waiting in line to register as undergraduate psychology students.

Sundali admired Rapoport the way that any student would idolize a professor who ushered him into a Wonka factory of ideas. They also had a closer relationship than most students and professors. When Sundali went on a skydiving trip with his college buddies, for example, Rapoport announced that he wanted in. "Which in and of itself was sort of remarkable, because he had a major heart condition and had been off for the semester getting his heart fixed," Sundali says. "I didn't know if I should be taking him skydiving." He took him anyway. The scholar piled into the car with two former college athletes who had barely graduated but still found gainful employment managing other people's money for august financial institutions. Sundali told his friends about a paper he'd read in Rapoport's class that claimed the hot hand was a myth, and they were barely on the highway before they commenced their grilling of the egghead. Rapoport took such a beating that he finally blurted out, "Is there *any* piece of evidence I could give you to show this isn't real?" Sundali's friends concluded there was no amount of evidence that would change their beliefs about the hot hand. "I'm glad you have faith, and I'm never going to discount someone's faith," Rapoport said. "If you have faith, evidence doesn't matter." The car pulled into the hangar and they were handed liability waiver forms. Rapoport gulped. This man who had suffered a series of heart attacks couldn't answer the health questions truthfully. He pulled Sundali aside.

"What should I do?" he asked.

"Well, if you want to go skydiving," Sundali said, "check yes and sign your name."

Rapoport ignored the evidence and went with faith.

The other reason that Sundali thought studying gambling would be worth his time was that his place of employment was the University of Nevada, Reno. There were as many casinos on the fringes of campus as there were coffee shops. When they set about publishing their series of papers about decision-making in gambling, Croson and Sundali looked at wagers made at the rapid roulette wheel in a Harrah's casino. "We got this huge stack of old IBM paper with the betting patterns of every player," Sundali says. "We know what they bet and when they bet it." Sundali later got a call from one of his students who worked for a manufacturer of slot machines and offered a mother lode of seventeen million slot pulls. "It's not the best game," Sundali says, "but it was the best data set." But they hit the real jackpot when one of Sundali's students in an executive MBA course told him that he worked for one of the local casinos. Sundali mentioned that he was about to send a Ph.D. student to the casino floor to record bets on a roulette table. But his student promised to do him one better. He could obtain the security videotapes from his casino's eye in the sky. "We had this Ph.D. student all excited," Croson says. "And now they're in a basement watching videotapes frame by frame and putting them into a spreadsheet." But now Sundali and Croson were also sitting on a gold mine: eighteen hours of an overhead shot of the same roulette table. They had 139 gamblers, 904 spins, and 24,131 bets at their disposal.

They started by looking at the fifty-fifty bets in roulette. Those were the wagers like red versus black—basically heads versus tails. There were 531 bets after gamblers had witnessed at least one spin of the wheel. They were split roughly down the middle between bets on the same outcome (red after red) and bets on the opposite outcome (black after red): 52 percent versus 48 percent.

But when Sundali and Croson analyzed the bets after streaks of two, the ratio suddenly flipped: 49 percent versus 51 percent. It tilted toward bets of the *opposite* outcome. That is, after the wheel landed on red, 48 percent of the next bets were on black. But after the wheel landed on red twice, 51 percent of the next bets were on black. That number climbed to 52 percent after three reds, 58 percent after four reds, 65 percent after five reds, and a whopping 85 percent after six or more reds. "They place significantly more bets against the streak than with the streak," Sundali and Croson wrote.

This was the corollary of the hot hand: the gambler's fallacy.

The gambler's fallacy has been infecting the brains of unsuspecting bettors for as long as they have been charitably donating their money to casinos. The first person to coin this phenomenon was a French mathematician, statistician, physicist, and god knows what else named Pierre-Simon Laplace. He was Kahneman and Tversky hundreds of years before Kahneman and Tversky. If he were alive today, Laplace would have his own podcast. He would be famous after giving a TED Talk. If he were really unlucky, he might even be invited to Davos. But he came to these conclusions a long time ago. In one chapter of his seminal 1814 book, *Philosophical Essay on Probabilities*, he used the example of the French lottery to describe the gambler's fallacy, noting how foolish it was that people favored the numbers that hadn't been picked in a while. "The past ought to have no influence upon the future," he wrote. This was especially true in casino games that stole people's money—games like roulette. For the same reasons that people were tricked by the lottery, people also overreacted to streaks in the casino.

The gambler's fallacy is different from the hot-hand fallacy. Stephen Curry makes three shots in a row, and everyone in the arena expects him to make a fourth. That's the hot-hand fallacy. But the roulette wheel lands on red three times in a row, and everyone in the casino puts their money on black. That's the gambler's fallacy.

The issue is how we perceive outcomes that we feel we control (basketball) and how we perceive outcomes that we know are beyond our control (roulette). When there's a streak that defies the odds, we understand that it will even out, and we place our wagers accordingly. We internalize the regression to the mean. But that's not what happens when *we* are the ones challenging probability. When it's Stephen Curry chasing that fourth consecutive shot, we believe he's temporarily beaten the statistical inevitability. He's *on fire*!

It's even possible for us to believe in both the hot-hand fallacy and the gambler's fallacy at the same time. Peter Ayton and Ilan Fischer came up with a clever way to illustrate this contradiction. They took some of their psychology students into a lecture hall and passed out a three-page quiz for extra credit. On each page was a jumble of @s and #s. It looked as if someone had butt-dialed a tweet.

As they tried to make sense of these seemingly incomprehensible sequences, the students were told that the symbols were the disguised results of six experiments their professors had conducted: basketball shots, coin flips, soccer goals, dice throws, tennis serves, and roulette spins. What the students weren't told was that those experiments never really happened. The series had been spit out by a computer that randomly generated sequences with eleven at signs and ten hashtags. Ayton and Fischer split their fake experiments into two categories. They pitted human performance (basketball shots, soccer goals, tennis serves) versus pure chance (coin flips, dice throws, and, yes, roulette spins). The students looked at twenty-eight sequences with different alternation rates between the at signs and hashtags. The visual differences were striking. This is how a series with a low alternation rate looked compared with a series with a high alternation rate:

Low: @ @ @ @ @ @ @ @ # @ # # # # # # # # # @
High: @ # # # @ @ # @ # @ @ # @ # @ # @ # @ # @

Ayton and Fischer said that each string of @s and #s could be makes or misses (as in basketball shots) or reds or blacks (as in roulette). And that's pretty much all the information their students were given. It was up to them to figure out which sequences represented what tasks, and the professors offered extra credit to any of their students who outperformed the class average.

By then Ayton had been thinking about the hot hand and gambler's fallacy for many years. One of the first things he'd written that anyone outside academia had bothered reading was a short essay for a popular science magazine in which he tried to replicate Gilovich, Tversky, and Vallone's basketball paper in soccer. After studying the goals of the English Premier League's top scorers, he reached the same conclusion. "Any belief in the 'hot foot' is also a fallacy," he wrote. He soon experienced the same ruckus. His essay caused such an uproar in his native England that he was invited on the radio to debate Ron Atkinson, a longtime soccer manager better known as Big Ron, about whether a hooligan would be right to believe in streaks. Big Ron had some nits to pick. "You've never been in the dressing room!" Big Ron yelled. "I've been in the dressing room. *I* know what it's like." Ayton was delighted about Big Ron's furor. He was getting precisely the same incredulous reception from soccer that Tversky had received from basketball.

But this particular experiment was different from most of Ayton's work because he couldn't predict beforehand if it would yield anything interesting about the hot hand or gambler's fallacy. "Often when you run an experiment, you're pretty certain what's going to happen, or at least you think you are, which makes you wonder why you're going to run the experiment," he says. "But with this one, I really didn't have a clue."

What happened next wouldn't have surprised Laplace. The students guessed the streaky sequences were basketball shots and the runs that seemed random were roulette spins. In their minds, @ @ @ @ @ was the product of human performance, but @ # # # @ was pure chance. When humans had control, they believed in the hot hand. When they were bystanders, they believed in the gambler's fallacy.

The roulette tables in Las Vegas are designed with those bystanders in mind. Right next to the wheels in today's casinos are electronic scoreboards that update after every spin with real-time statistics: the red versus black split, the distribution of numbers, and, most important, the last twenty numbers. It's an impressive display that is entirely meaningless. The ostentatious screens are as helpful to gamblers as a livestream of the Bellagio fountains. "There's no reason to put up a board with the prior numbers if it's a completely random game," Sundali says. "But they know the bettors care about the prior numbers."

They care because they believe there are patterns to be cracked. The casinos are targeting clients who believe in the gambler's fallacy. They're providing just enough information for those people suffer-

ing from this particular bias of the mind to make regrettable decisions. Those electronic scoreboards might as well be neon billboards
that scream: PLEASE GIVE US ALL OF YOUR MONEY.

That brings us to the next thing that Croson and Sundali studied:
Did the gamblers who embodied the gambler's fallacy also believe in
the hot hand? They did. When those people playing games of chance
thought they had the hot hand, they didn't leave the table until they
cooled off. They ordered the same drinks and performed the same
lucky rituals as long as they kept winning. They did everything in
their power to keep playing for the reasons that Stephen Curry tries
to keep shooting: because they're hot.

Croson and Sundali looked at the gamblers at this particular roulette table in this particular casino and found that 80 percent of their
subjects walked away from the table after losing, but only 20 percent
voluntarily stopped playing after a winning spin. You might think it
makes sense to leave after winning and quit while ahead. This is roulette! You're going to lose eventually. It would be smart to take your
profit and splurge on a steak dinner. Maybe that's what you believe.
But the casinos know it's not how you behave.

This wasn't the only evidence of the hot hand they uncovered.
Croson and Sundali also learned that if gamblers won fifty-fifty bets,
they became *more* aggressive. The roulette players who won their
fifty-fifty bets spread their chips over fourteen numbers on the next
spin.

The unlucky ones who lost made only nine bets the next time
around. The last thing that Croson and Sundali found when they
replayed the videos of three nights at the roulette table was their
coolest discovery yet. It was one thing to determine that the average
roulette player believed in the hot-hand fallacy and the gambler's fallacy. But what about the individual roulette player? It turns out the
people who bet according to the gambler's fallacy *were the same ones*
who bet according to the hot-hand fallacy. That is, if you bet on red

after black, you also probably bet more after winning a bet. "People seem to believe that people can get 'hot,'" Ayton and Fischer had concluded, "but that inanimate devices cannot." Croson and Sundali also emphasized that phrase: "*seem* to believe." "We don't actually know their beliefs," Sundali says. "All we're measuring is their behavior." That was the point of bringing their experiment to the casino. They could evaluate what their subjects did instead of what they said they would do.

But what if we extend the principles of the gambler's fallacy beyond the neon lights of the casino? What if they apply to baseball umpires, too? And what if people with jobs where the stakes are greater than calling balls and strikes also behave according to the gambler's fallacy?

What if that bias is what determines whether a famous Iraqi sculptor gets asylum in the United States?

4.

The paper was called "Refugee Roulette." It was the most comprehensive study of U.S. immigration ever published, based on more than four hundred thousand asylum cases, including those overseen by the people entrusted with making the single most important decision of another person's life: judges. It was an explosive piece of scholarship, and the law professors who wrote the paper made sure it would be read as widely as possible by giving it that name. The conclusion of their analysis of asylum applicants like Alaa Al-Saffar was a powerful repudiation of the way that justice is supposed to work. "Whether the asylum applicant is able to live safely in the United States or is deported to a country in which he claims to fear persecution," the authors wrote, "is very seriously influenced by a spin of the wheel of chance."

They were right. And yet they had no idea how right they really were.

The upshot of their paper was that immigrants subject themselves to that wheel of fortune as soon as they apply for asylum. Their odds of staying in the United States shift based on circumstances beyond their control. There is no such thing as a level playing field. In fact the same application could easily produce different outcomes. Chinese asylum seekers win 7 percent of their cases in Atlanta and 76 percent of their cases in Orlando. One immigration judge grants asylum in 6 percent of cases. Another immigration judge in the same courthouse grants asylum in 91 percent of cases. In cities like Miami, New York, and Los Angeles, 32 percent of the judges deviate wildly from the average rates. The outliers aren't the exceptions. They're the expectation.

"There is remarkable variation in decision-making from one official to the next, from one office to the next, from one region to the next, from one Court of Appeals to the next, and from one year to the next," the researchers wrote.

The most startling takeaway of this disturbing study was that nothing in a case was more important than the judge—not the asylum applicants, not the country they were coming from, not the skills they would bring to the United States, and not even why they were fleeing in the first place. There is a great deal of randomness that goes into assigning judges. A deserving immigrant who could do as much for America as America does for him might very well get denied asylum simply because he got stuck with a strict judge in Atlanta instead of a lenient one in Orlando.

But the process is especially cruel to asylum hopefuls because it's not only about the who and where. It's also about the *when*.

Their chances of staying in the United States depend on whether the judge who was randomly assigned their case recently granted asylum in a completely unrelated case. That's how arbitrary it can be. The asylum court is basically a casino.

Remember those economists who wrote that paper about baseball umpires? Baseball umpires weren't the only people they studied for signs of the gambler's fallacy. They also studied the decision-making habits of a collection of experts whose decisions actually mattered. They studied asylum judges.

Bruce Einhorn was one of their subjects. Before he was a U.S. Immigration Court judge, Einhorn worked for the Department of Justice, where he quite literally helped write the nation's asylum law. But first he was an undergraduate student at Columbia University, curious about the sort of psychology that can shape a judge's supposedly impartial judgments. "I find that stuff fascinating," he says. "I don't think all judges do." Einhorn trained laboratory rats in the Skinner box and studied their behavior. He ended up learning something about himself along the way. "I learned that I don't want to be a rat in a box," he says.

That much he proved in his two decades on the bench. Einhorn was a generous judge. He had a higher asylum grant rate than most of his peers. He did not say yes to everyone who came before him, but he said yes enough that it became clear that his default response was not to say no. "It's always easier to deny relief than to grant it," he says. There were a surprising number of incentives to saying no. Anyone who said yes as often as Einhorn risked putting himself in the crosshairs of bureaucracy. "You're facing the possibility that you're known as a wimp," he says, "instead of being known as a person of integrity who perceives himself to be what he is: a judge."

That even the people tasked with the authority of federal judges could look at the same cases and come to a wide variety of conclusions made an economist named Kelly Shue wonder if they were also prone to cognitive biases. She mentioned to her colleagues Toby Moskowitz and Daniel Chen that she had some data pertaining to asylum judges. She wanted to see if these real people in the real world with real stakes were subject to the gambler's fallacy. The caseloads are completely random, and judges are encouraged to complete as many as possible in as little time as possible, which makes the asylum court a breeding ground for bias.

What happens in one asylum case theoretically should have no effect on what happens in the next case. The problem with theory is that it doesn't account for the whims of people wearing robes to work. The judges are not robots. They are human beings like Bill Miller, the Major League Baseball umpire.

Judges occupy a complicated place in public society. We give them the authority to play god. They are our chosen enforcers of the natural order. But when they feel it's their job to balance things out, they become the opposite of the basketball fans cheering for a streak to continue. It is the judge's obligation to ignore the streak and make an impartial call. They often fail to uphold that standard. They notice the streak. They believe there shouldn't be a streak. So they end the streak.

Shue, Chen, and Moskowitz analyzed more than 150,000 decisions from 357 asylum judges. They calculated the average grant rate to be 29 percent. But when they looked at the sequence of cases, they could identify when the average shifted. And they found that judges were less likely to grant asylum *immediately after* they granted asylum in their last case. That is crushing. It means that an immigrant in need of asylum—someone who has already survived terrible hardship, lived through unimaginable misery, and outlasted awful odds—is automatically penalized for something that had nothing to do with the application. The depressing thing is that the statistics get more depressing. If a judge has granted asylum in two straight cases, he is 5.5 percentage points less likely to give asylum in the next case than if he'd denied two straight cases, regardless of the merits of the applicant. The judges were as vulnerable to the gambler's fallacy as the baseball umpires and the drunk roulette players. And someone like Alaa Al-Saffar could be screwed by this spin of the wheel.

"The judges understand that cases are brought to them randomly, but as soon as you say random, we have a really strange sense of what random means," Moskowitz says. "Most of us don't understand it very well. They think that random means I have six cases today, and half should be positive and half should be negative. They don't think you can get three in a row of the same type followed by three in a row of a different type. They think it should be alternating. And it's just not."

If anyone had bothered asking him to fix the system, Moskowitz would have proposed one simple tweak: change the way that asylum judges get assigned cases. He thinks every asylum applicant should be heard twice. "The judge makes the decision, and there's another judge who reviews the decision," Moskowitz says. "But they get to see the cases in a different order." That would eliminate the issue of two judges looking at the same application and seeing different cases simply because one of them has just granted asylum and the other has just denied asylum. It isn't the most practical solution, considering the judges already have far too many cases on their dockets without doubling their workload, but it is the right one.

Moskowitz likes this idea so much that he's implemented the behavioral adjustment in his own life. When his teaching assistants graded exams, he made sure that each one was graded twice, at which point he averaged the scores together. But after brainstorming ways to beat the gambler's fallacy, Moskowitz realized the flaw in his own system. "There could be some bias there," he says. "You see a couple great exams in a row, and it might influence what you do on the next exam." He took his own advice. His teaching assistants are now given the same exams to grade but in a different order. "That," he says, "is a pretty fair and unbiased assessment."

Alaa Al-Saffar wouldn't have such a luxury. There would be several variables in the complicated equation of his asylum: which judge he got, where he got him, and when he got him. That was the human effect of the gambler's fallacy. The judges weren't betting on black or red, and they weren't calling balls or strikes. They were deciding whether he lived or died.

5.

Alaa Al-Saffar moved into a community for senior citizens in Southern California, a few turns off a road called Avocado Avenue, not long after he applied for asylum in the U.S. He converted his garage into a cramped artist's studio and got to work.

His relocation had awakened the creative energy inside of him. "I was exploding," he says. The cheapest means of expressing himself was painting. He produced a series called *Lovely Dancers,* a collection of buxom women in various states of undress. "In my country, if I do this, they kill me," he says. But once he was allowed to do anything he wanted, he wanted to do something that he'd never been allowed to do.

Al-Saffar was settling into his unlikely, vaguely American life when I visited him on a blazing summer afternoon. I had read a story in the *San Diego Union-Tribune* about him, and now I was sitting across from him. He greeted me with a bottle of water and two Starbucks Frappuccinos, vanilla and mocha, insisting that I choose one. He wore the artist's uniform of loose black T-shirt, black pants, and socks with sandals. He smelled of stale cigarettes. He sat behind a table with an ashtray, scattered pencils, and a laptop playing highlights from that day's soccer matches. He barely had to stand to get to his easel. There were artifacts from his old life scattered around his workspace, and he dusted off a few shoeboxes with his press clippings, his stamp that won a national contest, and photos of his sculptures in all their glory. He even removed a thick binder from the end of his bookshelf and beamed with pride as he showed me his secret vice: a hidden stash of ouzo and arak.

He spent most of his time in this studio connected to his home. He'd been offered a proper workspace at a nearby college, but he said the gas would be too expensive. When his friends promised to drive him, he begrudgingly declined. "It's too much," he said. "That's a place for a younger man." Al-Saffar liked working from home. That was how he preferred to work in Baghdad, too. The proximity was crucial to him considering he was still always thinking about work. To him there was no such thing as work-life balance. "It's my life," he said. "Without work, there is no life."

But there were two issues with his work and his new life. The first problem was the rocks. When he visited his local Blick Art Materials shop, he was surprised to learn that even the smallest hunks of sculpting material stretched his meager budget. The second problem was one that he couldn't solve by painting instead of sculpting.

Al-Saffar had been in Southern California for several years already, and he still couldn't say with any confidence if he would be allowed to stay. He'd left his old life and sought a new life at the exact moment in American history when the people who deserved compassion were treated with disdain. He couldn't escape his old life because his new life was now at the mercy of the U.S. asylum system.

It was a system that required Al-Saffar to scale a small mountain of paperwork before he could even formally apply for asylum. The application contained such tedious instructions that anyone who finishes reading them should be granted citizenship on the spot. There was a purpose to all this paperwork. It was meant to suss out who was serious about needing to be saved. Al-Saffar had to demonstrate a credible fear of death back in Iraq and prove that he was under persecution for his race, religion, nationality, politics, or, in his case, membership in a particular social group. He recalled the threatening letter under the garage door that his daughter found and explained he was an artist who feared harm if he were to return to Iraq.

The next step in the asylum process was an interview with a

Department of Homeland Security asylum officer whose job it was to put a face to all that paperwork. Al-Saffar was granted his first interview a few months after he applied. It seemed fast, all things bureaucratic considered, and his family was elated. This was surely a good sign. Once he was in a room with an officer, Al-Saffar attempted to explain all the reasons he was seeking asylum. "I want to do something here—for America," the Iraqi sculptor said. It was a heartwarming gesture that would have little effect on his application. Instead he would have to convince the government that his motives for staying in the United States went beyond his burning desire to make art. He would be allowed to stay only by arguing persuasively that there wasn't another choice. Al-Saffar had to prove that he was petrified. The asylum officer charged with establishing the credibility of his fears needed to know more about Al-Saffar's past, which meant he needed to know whether he knew Saddam Hussein.

I am poor man, Al-Saffar recalled saying. *I am artist.*

Did he work for Saddam Hussein?

Yes.

What kind of work?

I am sculptor, he pleaded.

Did he feel in danger when he was in Iraq?

In my country, I need five eyes: one, two, three, four, five, he said, pointing to the left side, right side, and back side of his head.

And how many eyes did he need in America?

Just one, he said.

The officer who interviewed Al-Saffar could have handled his application in several ways. He could have granted him asylum on the spot. He could have denied him asylum. Or he could kick his case down the road and refer him to a federal immigration judge to decide whether he deserved asylum.

It had been years since that interview and months since his last

contact with the asylum office when I found myself drinking Frappuccinos on that hot summer day with Al-Saffar. The U.S. asylum
system was in crisis. The wave of migrants from Central American
countries flooding the border had overwhelmed the officers and
immigration judges responsible for the backlog of applications. The
immigration system would have collapsed under the weight of a 100
percent or 200 percent or 300 percent increase, but what actually
happened in the years before Al-Saffar applied for asylum was a catastrophic deluge. The real number of asylum cases had increased over
the course of five years by *1,750 percent*. The immensity of that figure
is impossible to comprehend at the scale of the U.S. immigration
system. So let's simplify it. Imagine you're in the habit of keeping a
daily to-do list. On a busy day, you have ten tasks. You can derive a
sense of satisfaction when you cross off every item. But an increase
like the one that disturbed the asylum system would amount to
ten chores becoming *185*. That's no longer a daily to-do list. It's a
monthly calendar packed into twenty-four hours.

There were 320,663 people like Alaa Al-Saffar with pending
cases, enough to fill four NFL stadiums, and the immigration courts
were overworked to the point where they weren't actually working.
The backlog kept growing. Meanwhile the government had adopted
a last-in, first-out approach in which the first applicants to be heard
would be the ones who arrived last. The agency was plowing through
asylum filings at such a glacial pace that it seemed like it would never
get to Al-Saffar's. With so many migrants at the border, it was easy
to ignore the asylum candidates who already had temporary visas,
the hopeful Americans like Al-Saffar. The month that I went to see
him ended with the affirmative asylum division of U.S. Citizenship
and Immigration Services clearing about seven thousand cases and
leaving hundreds of thousands on the docket. There were some who
were granted or denied asylum on the spot, but the great majority of
the applicants were interviewed and then referred to federal judges.

They joined Al-Saffar in asylum purgatory. "Each way, you are dead," he says. "Sleep or awake."

During the years that his application was on hold, his family regularly called the local asylum office for an update. They were tortured with the same response each time. "Pending, pending, pending," Zinah says. The whole ordeal had been so anguishing that she was no longer sure she'd made the right decision encouraging her father to leave his old life in Iraq and seek a new life in the United States. That was how dreadful the asylum process could be: Alaa Al-Saffar's own daughter thought he might be happier in a place where there was a good chance that he would be killed for being an artist. "I wanted him to stay here for his safety," she says. "But now, believe it or not, after all this waiting, the case is still pending. I regret it. Did I do the right thing by keeping him here? I don't know. He for sure misses home."

Al-Saffar liked to walk around his new home to clear his mind of this existential torment. He waved to his neighbors with American flags in their rock gardens. He was a popular guy around the community. His friends rushed over to Zinah to say hello whenever she visited. "I want people to know how talented he is," she says. "How sad it is that talented people don't get recognized. I want him to get recognized before his time is over."

But there was one more thing that Zinah wanted for her father.

"I want him to have a choice in life and a choice in art," she says. "Not to always be dictated by somebody."

Alaa Al-Saffar passed the time as he waited for the results of his asylum application the only way he knew how. He continued to make art. His goal was to build a monumental public sculpture in the middle of his American town. He wanted to give back to the place he was still hoping would take him in. He even made an immaculate little model of his magnum opus. It was a map of the United States combined with the American flag, held together by a word in English and repeated in the many languages of the famous Iraqi sculptor's native land: "freedom." He couldn't be sure that he would get to experience it for himself.

Alaa Al-Saffar's uncertainty. Jed Lowrie's home run. Nick Hagen's bumper crop and David Booth's rules of basketball. Rob Reiner's movies, Rebecca Clarke's sonatas, and William Shakespeare's plays. Mark Turmell's addictive game and, of course, Stephen Curry's shooting extravaganza.

By now we know what the hot hand looks like. But there is one thing we still don't know.

Should we believe it when we see it?

Six

THE FOG

"Be true to the data."

1.

As the biggest city in a politically neutral country, Stockholm was an exceptionally strange place to experience World War II. It was close enough to battlefields that it was a safe haven for spies and diplomats but far enough that everyday life there felt almost normal. "To a traveler who has seen bomb-pitted Britain," one foreign correspondent wrote, "Stockholm seems a little bit of heaven." Swedes still went to concerts and the opera. They still went dancing in tuxedos and gowns. And they still went to the movies.

In the bitter winter of 1942, the hottest movie in Stockholm was *"Pimpernel" Smith,* a film about a swashbuckling professor on a covert mission to free concentration camp prisoners. It was banned in Sweden for being too political, but the fact that it was banned didn't mean it was hard to find. This censorship actually had the opposite effect. It only made even more people want to see it. To keep up with the demand, the British embassy hosted a screening of *"Pimpernel" Smith,* and one of the people who came to watch a hero battle evil was a young man named Raoul Wallenberg. He walked out that night feeling overwhelmed by inspiration.

"That," he said, "is something I would like to do."

But he was never supposed to do anything like *that*. Raoul Wallenberg belonged to a family that was the Rockefellers of Sweden. TO BE—NOT TO BE SEEN was the motto on their crest. Raoul was always different from other Wallenbergs, though. He never enjoyed the comforts of aristocracy. His father died when his mother was pregnant with him, and he made a deathbed wish for the son he would never meet: "I will be happy if only little Baby becomes a nice and good and simple person." Raoul was nice and good. He was anything but simple. He was groomed from a young age by his paternal grandfather, a Swedish ambassador who understood the importance of being worldly and empathetic, which was the reason he shipped Raoul to the University of Michigan to study in a wholesome American environment. "The conviction here at home that we are better than anyone else needs to be shaken," he wrote in a letter to his grandson, later adding, "A trailblazer discovers the good to be found out there among the foreigners."

Wallenberg's time out there among the foreigners left him idealistic and wildly ambitious. He knew he wouldn't be in the United States forever and drenched himself in Americana while he could. Wallenberg was unpretentious about his travels around this foreign country. He derived a special pleasure from hitchhiking, and he took pains to assure his grandfather that it wasn't beneath someone of his family's stature to ask random strangers for a ride. "Hitchhiking gives you training in diplomacy and tact," he wrote in a letter. "As for the risks, they're probably exaggerated." But not always. He was robbed at gunpoint one night while hitching back from Chicago and surprised himself when he found the whole thing unusually thrilling. He turned out to be so cool under pressure that it unnerved his assailants.

What he couldn't have possibly realized was that he was getting valuable experience training for another job—one that was far more daring than anything his grandfather had in mind.

After he graduated with a degree in architecture, Wallenberg kept traveling the world, and he continued to immerse himself in the local cultures wherever he went. As he sharpened his keen sense of place wandering South Africa and the land that would one day be Israel, he developed a fondness for the Jewish people he encountered along the way. He admired them so much that he started bragging about one of his distant relatives, his maternal grandmother's paternal grandfather, who technically made Raoul Wallenberg one-sixteenth Jewish.

As he spent more time out there among the foreigners, he began to think he didn't want to follow the path his grandfather had set for him. He wouldn't allow himself to be another Swedish banker. He couldn't "sit around saying no to people," as he once put it. Wallenberg had bigger aspirations. He wanted to leave behind a legacy beyond his family crest. He craved a purpose. And then a purpose happened to present itself.

When he finally moved back to Sweden, Wallenberg found himself in the business of buying goose meat. His first assignment was to visit Budapest. He came back with disturbing news. For one thing, he didn't have fifty tons of goose meat, which had been the reason for his visit in the first place. But the more important takeaway was that something appeared to be deeply wrong in Hungary. The spread of anti-Semitism was obvious to anyone who bothered paying attention. One of those people was a fur trader named Norbert Masur. He was troubled enough by the situation in Europe that in April 1944, one month before the first of more than six hundred thousand Hungarians would be deported to death camps, he wrote a letter to his local rabbi. "We should find a person, highly skilled, of good reputation, a non-Jew, who is willing to travel to Romania/Hungary in order to lead a rescue mission for the Jews," Masur wrote.

The word of Masur's letter reached Wallenberg. *This* was the job for which his grandfather had inadvertently groomed him. He applied to be a hero.

At almost exactly the same time, the War Refugee Board of the United States sent urgent telegrams to embassies in five neutral countries looking for the type of person Masur was seeking. The Americans wanted as many diplomats as possible in Hungary as quickly as possible. Sweden was the only country that was interested in helping. And there was only one Swede for the job.

Raoul Wallenberg was officially named a Swedish diplomat a few days after the Allies landed on the beaches of Normandy in June 1944. No one seemed to care that he wasn't only working for Sweden or that he wasn't technically a diplomat.

His life as he knew it was over from the minute he stuffed a windbreaker, a sleeping bag, and a revolver into his knapsack. Wallenberg was in a rush. He couldn't bear the idea of being in Sweden for any longer. "Every day costs human lives," he told his superiors. His train rumbled into Budapest in July 1944, and he was immediately briefed on the humanitarian efforts already under way.

Wallenberg learned that the Swedish embassy had been distributing official documents to protect Jews from persecution, and these provisional Swedish passports spared people from internment, deportation, and the safety risks of wearing yellow stars to announce themselves as Jewish. As long as they had official Swedish documents, they could identify themselves as neutral Swedish citizens. The passports had been mildly successful, but there weren't nearly enough of them, only hundreds for hundreds of thousands of endangered Jews. "I think I've got an idea for a new and maybe more effective document," Wallenberg said.

He called it the Schutzpass (of course the Germans had a word for "protective passport"), and his rule was that anyone in possession of a Schutzpass would be fully protected by the diplomatic powers of neutral Sweden. At first he made 1,500. And then 2,500. And then 4,500. And then so many that he stopped counting.

The early Schutzpasses were reserved for Hungarian Jews with legitimate Swedish connections, the kinds of people who might plausibly have such papers, but soon the demand exceeded the supply to the point there was a black market for the passports. There were two things that made this development all the more remarkable. The first was that the Schutzpasses were free. The second was that the Schutzpasses were fake. "Not even the name Sweden was written in Swedish on what claimed to be an official Swedish document," wrote Ingrid Carlberg, one of Wallenberg's biographers. There was nothing even remotely official about this document. In this sense the Schutzpass was like Wallenberg himself: it looked and acted the part, and everyone played along. Wallenberg's allies were amazed the Nazis could be so impressed by something that was so fake. But that was the one convenient thing about fighting a war against people attracted to evil causes: a great many of them were fools. They could be easily duped by anything that implied authority.

It wasn't long before Wallenberg unilaterally declared that the application process for Schutzpasses had become too burdensome. The Nazis were brutally efficient killing machines. Their accelerating genocide left him no time for diplomatic niceties. And because he wasn't a diplomat, he wasn't interested in being nice anyway. He issued orders to approve every Jew who applied for Swedish papers even if he couldn't find Sweden on a map. Wallenberg had been right about himself all those years ago. He did have the potential to do something more than say no to people. "If anyone is capable of making their way to our door and submitting an application for a protective passport," he told his staff, "the answer from now on will always be yes."

That decision was how Raoul Wallenberg turned out to be one of the greatest heroes in the history of mankind. What he accomplished in the next six months was nothing short of superhuman. There are few people who have ever done so much good in so many ways in so little time.

At this point you might be thinking, *What in the good name of goose meat does Raoul Wallenberg have to do with the hot hand?* We'll get to that.

But first let's marvel at his many acts of bravery. He plunged into icy waters to rescue Jews as Nazis were shooting at them. He packed twenty thousand people into safe houses that were designed for five thousand people. He pulled Jews off trains bound for death camps with his bare hands. He used all the skills he'd collected over the course of his eclectic life and some that he wasn't aware he possessed. He was so tireless that he slept only four hours a night. He was so charming that he managed to befriend the wife of Hungary's Far Right foreign minister. He was so persuasive that she gave her husband an ultimatum one night at the dinner table: honor Wallenberg's protective passports or lose her forever. He hurled china at her, claimed betrayal, stormed out, but eventually agreed to her demands.

It seems impossible that one person could have pulled any of this off in a lifetime. Wallenberg did all this in a few months. He saved approximately one hundred thousand human lives with the power of his imagination.

The culmination of his period of unimaginable courage was a rather surreal confrontation with his rival. One night near the end of the war, at the height of their conflict, Wallenberg had dinner with the Nazi mastermind Adolf Eichmann. Eichmann was obsessed with Wallenberg. He referred to him as "that Jew-dog Wallenberg"—as in "have that Jew-dog Wallenberg shot." On one side of the dinner table was a man who represented the banality of evil. On the other side was a man who personified the ingenuity of good.

They put aside their differences to eat and drink brandy together, and they retreated to the living room when it was time for coffee. Wallenberg opened the curtains to look outside. The night sky was red. They could see artillery fire as the Russians inched toward Budapest. It was at this point when Wallenberg felt it was the right time to tell Eichmann that the Nazis would never win the war. "I admit that you are right," Eichmann said, to the shock of everyone in the room. But only after abandoning the very cause that he embodied did Eichmann put a chill on the evening with an ominous warning to Wallenberg. "Accidents do happen," Eichmann said. "Even to a neutral diplomat." There was no need for any more coffee. Wallenberg couldn't feign the coolness that he felt upon being robbed while hitchhiking to Ann Arbor. But even if he was terrified, the only way he knew to cope was to keep working around the clock. "Of course it gets a little scary sometimes," he told one of his colleagues. "But for me there's no choice."

The Red Army sieged Budapest a few days after their dinner, and Wallenberg spied an opportunity in this shifting of power. He'd dreamed about helping rebuild the city once the worst was over. It was a reconstruction that seemed possible in the same way that everything he'd achieved in the last six months had been. When he was awakened early one morning in January 1945 to the sound of Russian street patrols, he demanded to speak with top Soviet authorities so he could pitch them his plan. Wallenberg expected to be gone for at least a week to meet with a Soviet military commander in Debrecen, Hungary. He left on January 17, the day after the liberation of Budapest's ghetto, and he couldn't shake the feeling that something was off. He was increasingly unsure if he should trust the armed Soviet officers assigned to be his ushers. "I do not know if I am a guest or a prisoner," he said.

What he also did not know was that the Soviet Army had secretly issued an order for his arrest.

They continued to assure him that he was a guest, not a prisoner, when his train left Budapest. As concerned as he must have been, he may have believed them. He distracted himself in his train car by working on a spy novel that he'd begun writing. But when the train pulled into Moscow, he was escorted to a building on Lubyanka Square. It was the headquarters of the Russian ministry of security. He thought he would be there for the night.

Raoul Wallenberg walked inside and disappeared forever.

2.

This chapter is not about Raoul Wallenberg. Not really, anyway. It's about the search for truth about the hot hand. It's about coming to smarter conclusions about what we know and don't know and think we should know but don't. It's about data, but it's not about bigger data. It's about *better* data.

Which brings us to another couple of Israelis who met in an extremely Israeli way. Gal Oz happened to know someone from the army who happened to be married to someone who happened to study with a professor who happened to be friendly with Miky Tamir. Tamir was more accomplished than Oz if only because Tamir was more accomplished than pretty much everyone in his orbit. He was a nuclear physicist who published scholarly papers about drones one day and drafted top secret classified documents the next. After several decades working for research centers and defense contractors, Tamir reinvented himself as a serial entrepreneur, and this was perhaps the most deeply Israeli thing he could've done. Israel was a country where not knowing someone with a start-up was about as likely as never having eaten hummus. By the time he met Oz, a young engineer in the Israeli Defense Forces, Tamir had successfully launched several companies and was starting to think about his next big idea. Oz thought he could help. He specialized in

Tamir's field of visual intelligence. They both knew how to use maps and data from satellite and aerial imagery to inform decisions. They both could see things before everyone around them. But what their shared expertise really meant was that once Oz spread the word he was leaving the military, it would only be a matter of time before he was introduced to Tamir. "Israel is small," Oz says. "Everybody knows each other."

Tamir wanted to bring all the futuristic technology from their field of visual intelligence to an industry that needed all the help it could get: sports. He didn't have to convince Oz of the potential. "I saw sports from the tech side," Oz says, "and it was more or less the same as what I did in the army." The origin story that persisted after they started SportVU—as in a different way to view sports—was that the company was based on proprietary Israeli missile-tracking technology. It wasn't true, but they didn't see the harm in letting the myth take on a life of its own, partly because it added to the intrigue about their company, but mostly because Israeli missile-tracking technology wasn't too different from what they were actually doing. "Tracking a missile is much easier than tracking a ball," Oz says. "A missile is much more predictable."

SportVU quickly caught the attention of another company on the cutting edge of sports called Sports Team Analysis and Tracking Systems—STATS for short. STATS executives believed that player tracking was the future of their business, but they didn't have the manpower to build such technology by themselves. They would have to acquire it. Before they spent millions of dollars to buy SportVU, however, they needed to know a whole lot more about the company. They asked a team that included a man named Brian Kopp to investigate.

A born-and-bred Midwesterner with a buzz cut, Kopp had never worked in sports before. He'd spent a few years in banking and private equity before he, too, decided that he didn't want to say no to people for a living. He enrolled in business school and took a job in strategic planning for an education company. "Because of course that's what you do before you get into player tracking in sports," he says. The next hairpin turn in his career brought him to STATS. He was still new on the job when his bosses told him to get on a plane to Israel and meet some guys named Miky and Gal who were supposedly applying missile-tracking technology to sports. "And so I went," he says. "I didn't know what the hell we were doing."

He read the book *Start-up Nation* on the plane to learn more about the entrepreneurial spirit of this country where he'd never been. But he was still unsure about whom he was meeting and why they were meeting when he touched down in the bustling tech hub of Tel Aviv in 2008. He still couldn't wrap his head around his own company's strategy. It was only apparent to him in retrospect.

"It was very simple," he says. "Can we use technology to collect data that no one else has access to?"

Tamir and Oz gave Kopp a tour of the SportVU office. It didn't take long: the SportVU office was one room. "It was just a couple of guys doing something interesting," Kopp says. But it was becoming clear that this company, which had only dabbled in soccer and had a name that reminded him of visiting the optometrist, was developing technology with the potential to radically transform professional sports. *This isn't fully baked yet*, Kopp thought. But there's something here. And we'll get it early. STATS took his advice and bought SportVU. It cost $18 million. Tamir took his cut and moved on to find the next next big thing. Oz's job was to build SportVU. Kopp's job was to sell SportVU to sports that weren't soccer. He knew exactly where to start: basketball.

What made basketball so popular was also what made it so difficult to quantify. It was a balletic exercise in outrageous athleticism. How could anyone put a number on the perpetual action taking place over 4,700 square feet of space with ten players and a leather ball in constant, intertwined, and unpredictable motion? Well, with an elaborate tracking system based on something not unlike Israeli missile-tracking technology. That was Kopp's pitch when he was invited to the NBA Finals in 2009 to present to the league's top executives. When he started thinking about how he could tell them about SportVU, he decided that he had to show them instead. SportVU's engineers flew to Orlando and positioned their cameras to record the action on the court down below. The game between the Los Angeles Lakers and Orlando Magic was on a Tuesday night. Their presentation was on Thursday afternoon. They pulled all-nighters to search for the one tidy example that would stupefy their audience.

They found it early in the first quarter with a play featuring the two largest people on the court. Andrew Bynum, the center for the Lakers, made a nifty spin move and tried a hook shot. Dwight Howard, the center for the Magic, came swooping across the lane and swatted the shot away. The referees had to make a snap decision: Was it a legal block or an illegal goaltend?

The question of whether the ball was already on the way down was almost impossible to discern in real time. The referees had milliseconds to make a judgment call that came down to millimeters. SportVU's employees had days to make an informed, objective decision based on robust data. Their cameras had been tracking the ball the whole time. Since they had the precise coordinates of its path, they could determine if the referees were right.

Those referees who were entrusted to officiate games in the NBA Finals had ascended to the top of their profession because of the way they had been trained to use their eyes, their intuition, and their own systems of pattern recognition to make calls. There had never been a better option. But now the referees didn't have to make educated guesses based on their years of experience. They didn't have to process the play while running it through a database in their minds with millions of similar plays. And they didn't have to do it in a matter of milliseconds. In the same way that players and teams would soon use SportVU to determine strategies, the referees could make decisions based on what the data told them, not what they happened to believe, because what they happened to believe had the potential to be shaped by all the biases they carried with them as human beings.

Kopp was about to show the NBA's senior executives that there was a better way. It was a means of injecting empiricism into a field rife with ambiguity. What he didn't know was if anyone would be willing to listen. He feared none of them would be at the presentation. He was surprised when all of them crowded into a cramped hallway that had been curtained off for the occasion. There were even some team officials who'd invited themselves to Orlando not for the Finals game that night but for the presentation from Kopp that afternoon. They watched as he rolled out a television monitor to replay Dwight Howard's possible goaltend of Andrew Bynum. The screen showed a digital re-creation of the ball with the x, y, and z coordinates of its arc toward the basket. SportVU calculated the height of the ball by measuring its distance from the ground on a frame-by-frame basis. The height of the ball was decreasing when Howard blocked Bynum's shot. It had already reached its apex. This meant the referees had made the right call: it was a goaltend.

SportVU would become so advanced in the coming years that Kopp is almost embarrassed about how rudimentary it was when he pitched the NBA. "It was like the *Pong* version of video games," he says. "What we showed was like *boop boop boop boop boop*." But the league's most powerful people were blown away. They wanted the NBA to be in business with SportVU.

The deal talks lasted for months even though the company didn't know what to charge or what it was providing and the league didn't know what to pay or what it was getting. SportVU was more of a promise than a real product at that point. "We were trying to negotiate the values of something that didn't exist," Kopp says. As they reached the finish line of all their haggling, the floating heads of NBA executives appeared on a videoconference screen in STATS headquarters for one more negotiating session, and Kopp could tell without any form of technology that he was in trouble. The suits at the league suddenly looked as if someone had reminded them they were about to pay huge sums of money for something that might not actually work. The deal was dead. Kopp's boss was furious. "You guys need to think about this and get back to us!" the CEO of the company bellowed.

Click.

STATS hung up on the NBA.

"They're going to call back," Kopp's boss predicted.

"They're not going to call back!" Kopp replied.

They did not call back.

A basketball junkie who'd played in high school and college, Kopp wasn't ready to give up on his NBA dream. If the league office was skeptical, he'd try working directly with teams instead. And if there was anyone who would get a special kick from making a deal with a start-up that hadn't been formally approved by the league office, it was Mark Cuban, a rebel who made a killing in the dot-com boom and turned his fortune into ownership of the Dallas Mavericks.

Cuban was hell-bent on winning an NBA championship and eagerly threw cash at every inefficiency he could find. Unlike free-wheeling baseball teams, basketball teams were limited by a salary cap. Since their payrolls were more or less the same, the advantages were on the margins, and Cuban flooded those margins with money. He invested in the biggest airplane and the nicest locker room to entice the most talented players to sign with his team. He also invested in the latest data-guzzling technology. He even hired grunts to code each game's location data for the Mavericks. He wasn't the only one to realize that just because something had always been done one way didn't mean there wasn't another, better way. The Houston Rockets

were outsourcing the same kind of work to be done by hand in In-
dia. But they couldn't do it for every team and every game. SportVU
could.

The Dallas Mavericks became the first major American sports
team to pay for SportVU when they had their arena equipped with
six high-resolution tracking cameras. They spent the bulk of the 2011
season trying to capture and quantify everything that happened
on the court. "We used so many different sources of data," Cuban
says. They built their own internal analytics and bought others like
SportVU. If there was any kind of data that had even a remote chance
of helping them win one more game, the Mavericks were interested,
and some of those metrics inspired them to save their most uncon-
ventional strategies for their most important games. In the NBA Finals that season, they decided their best chance of stopping LeBron
James was a guard named J. J. Barea, who was generously listed at six
feet and might have been that tall in lifts. One of the shortest players
in the entire league made an enormous difference for the Mavericks.
The season that began with the installation of Kopp's cameras ended
with the NBA championship.

SportVU was here to stay. The first six teams to buy SportVU were
the Mavericks, Rockets, Boston Celtics, Oklahoma City Thunder,
San Antonio Spurs, and none other than the Golden State Warriors.
The only remaining question was how much they could afford to
spend. If you divided the amount each team spent on payroll by its
number of wins, the price of a single win in the NBA was about $2
million. Since it was obvious that SportVU would help NBA teams
win at least one game, that should have put its value somewhere
north of $2 million per year. But the budgets of NBA teams didn't
match their brains. This wasn't the same as spending on a player. It
was more like an office expense. SportVU was the toner cartridge
in the printer. The smart NBA teams saw the $30,000 price tag as
a truly extraordinary bargain. Here was a tool that barely cost any-

thing but could turn a basketball game into a collection of data rich enough to alter the behavior of millionaires competing in a zero-sum marketplace.

But immediately after the Mavericks won the championship, there was an NBA lockout. The business of basketball stopped for six months. At first this was a welcome development for Brian Kopp. NBA teams suddenly had nothing better to do than listen to him. When he visited the New York Knicks, he figured they would bring a few people to a meeting that would last a few minutes. They brought their entire staff to a meeting that lasted three hours. "The lockout was one of the best things that could have happened to us," he says. "It opened the eyes of people who might have been resistant to the use of data. It almost made them have to pay attention."

The problem for Kopp was that his bosses were paying attention, too. They had gambled $18 million on an Israeli start-up, but all they had to show for their investment three years later were some measly $30,000 checks and one season of data.

It was a maddening time for Kopp. He felt that he and the rest of the STATS crew hadn't even begun to scratch the surface of possibilities with SportVU. They were only asking questions. They were still waiting for answers. His dream was to hire a team of mythical geeks who could delve into the SportVU data and report back on what they encountered. But he was also a pragmatist. He could read a balance sheet. And he didn't have to go to business school to understand that his budget wouldn't allow him to poach the analysts of his fantasies. Kopp was stumped. Ever since his first trip to Tel Aviv, he'd sensed that he was on to something and it was something big, something that could influence the way sports were played forever. It no longer sounded preposterous when he said it aloud. The whole point of Kopp's job was to reduce uncertainty. Now he was surrounded by it.

What he didn't realize was that the mythical geeks of his imagination already existed and he wouldn't have to pay them a dollar. They had dependable jobs, fancy titles, and reliable sources of income. The idea of making any money from SportVU never even crossed their minds. The data he was offering was so rich they might have paid him. And they weren't "mythical geeks." They were professors.

These researchers with sterling credentials were the saviors of SportVU. They could see the potential of this technology without having to sit through a formal presentation. These were smart people with sharp questions, and they were so exuberant to apply the scientific process to basketball that many of them already had their hypotheses ready for testing.

There were certain things they had always believed to be true. But they could never know for sure. They didn't have the right data.

3.

The war had been over for more than a year when Maj von Dardel wrote the letter that she hoped would end her personal hell.

"Dear Mrs. Roosevelt," she began.

Her wrenching, typewritten note to Eleanor Roosevelt began with a formal introduction and a plea. "Knowing your warmheartedness and kindness to all those who suffer, I have gathered courage to write to you," she wrote. "I am the Mother of the Swedish Secretary of Legation Raoul Wallenberg of whom you may have heard."

She continued: "As a mother, I am no unchallengeable witness but I know that his coworkers and the people he saved could all tell you about his remarkable courage and ability, which enabled him to risk his life day after day in this gamble with armed criminals with the lives of thousands of innocent people at stake. . . . The fact that a great part of the Hungarian Jews have survived can be attributed essentially to one man working as the representative of the Swedish king and the American president—my son."

Maj von Dardel had reached out to the former First Lady of the United States because her son who had saved so many lives needed someone to save his. Raoul Wallenberg hadn't been a free man since he walked inside the secret police headquarters on Lubyanka Square. The exact reason for his arrest and imprisonment remains unclear to this day, and everyone following his ordeal would be haunted by a question that has never been answered to anyone's satisfaction: What happened to Raoul Wallenberg?

Many years after his mother wrote to Eleanor Roosevelt, she wrote another heartbreaking letter to the Nazi hunter Simon Wiesenthal, the man who helped capture Eichmann himself. "Not knowing is the worst," she said. "To know that my son may be alive, that he may be suffering, has been admitted to a mental institution, is starving in a prison or forced to perform hard labor is much worse than if I could know with certainty that he were dead."

But the tale of Wallenberg's vanishing had devolved into a maze of lies, contradictions, and deception. By then the Soviet Union had offered several conflicting accounts of his fate. The obfuscation was enough to make anyone's head spin. First officials said that Wallenberg was in Russia and safely under Soviet protection. Then they said that he'd been murdered. Then they said that he was not in the Soviet Union and never had been and that no one even knew who he was. Then they said he'd died of a heart attack. And then, finally, they said he'd been executed.

The stories of Wallenberg's heroism were beginning to spread around Sweden, however, and his twist of fate brought yet more nobility to his family's name. But the name still meant nothing to most people in the United States when Marvin Makinen heard it for the first time. He would become familiar with the Wallenberg case under circumstances that he would never forget.

Makinen was on the path to medical school when he decided in 1960 to spend a year abroad in Berlin. It turned out to be more than a year when he was approached in May 1961 by two U.S. intelligence officers who asked him to be a spy. He agreed to their request. Makinen rented a green Volkswagen Beetle and drove toward the Soviet Union as a tourist. He was as much of a tourist as Wallenberg had been a diplomat. Makinen's mission was to snap covert photographs of military facilities on his way to and in the Soviet Union. He was taking pictures of barracks in the outskirts of Kiev when he was arrested by KGB agents on suspicion of espionage. Makinen was held in solitary confinement for three months, and he was found guilty by a military tribunal and sentenced to two years in prison and six years in a labor camp. Makinen was supposed to be going to medical school. Now he was going to a Soviet prison.

He spent the next twenty months in a prison several hours outside Moscow in a city called Vladimir. It was the place where the most notorious political inmates were incarcerated, and the conditions inside Vladimir Central Prison were as bleak as you might imagine. His saving grace was that he didn't have to serve his full sentence. Makinen was freed as part of a spy trade—two Americans for two Russians. The other American was a white-haired priest. At one point he'd been declared legally dead by the Vatican. But when they walked off the plane, he looked more alive than Makinen. Makinen had weighed 155 pounds before prison. By the time he stepped on a scale in the United States, he had already been home for a week, and he could tell that he'd added several pounds to his emaciated frame. When he finally looked down, he couldn't believe the number staring back at him. He weighed 105 pounds.

It goes without saying that Vladimir was not the easiest place to make lasting friendships. Even the most basic communications required the prisoners to be resourceful. They developed an elaborate tapping system, passed notes to each other, and traded gossip when they were transferred from cell to cell. But those brief conversations were valuable sources of intelligence. They were how Makinen came to possess one of the few things that he took with him back to the United States. It was the rumor that somewhere inside the Vladimir Central Prison was a prisoner from Sweden.

"I always thought that was odd," he says.

Sweden had remained a neutral country. The idea of one of its citizens being in a Soviet prison *was* odd. While he was debriefed in the State Department, Makinen mentioned this rumor about a Swedish prisoner, someone named Vandenberg. Then he was invited to the Swedish embassy a year later. Makinen was puzzled. He'd already been interviewed by the Swedes. Why did they need to speak with him again? His confusion gnawed at him until he was saying his goodbyes after that second interview and stumbled upon a clue. He told one of the Swedish diplomats that he was going on a date that night with a Swedish exchange student. It was true, but that's not why he said it. He said it to provoke a reaction. He got what he was looking for. "We ask you not to talk to anyone about this," the Swedish diplomat said. That response stuck with Marvin Makinen.

Who was *this person?* he thought.

They wouldn't tell him that much. But when he asked what this person did, they told Makinen that he'd been arrested in Budapest "helping Jewish people to escape from the Nazis," he recalls.

It would be another sixteen years before he found out more.

By then he was a professor in the University of Chicago's department of biochemistry and molecular biology. He'd survived one of the most stressful experiences any living man had endured and now he was a respected professor. Makinen came home late one night in 1980 after a long day at his lab conducting spectroscopic studies requiring liquid helium. It was about three A.M., but he was still wired from work, and he poured himself a glass of orange juice and took the *New York Times Magazine* to his living room. As he flipped through the pages, he came across one story that captured his attention. It was called "The Lost Hero of the Holocaust." He nearly choked on his orange juice once he started reading.

For almost two decades, he'd believed there was a Swedish prisoner in the Soviet gulag, someone named Vandenberg. Now it all made sense. Makinen realized why the Swedish embassy was so interested in his testimony. It was because he was half right. There was a Swedish prisoner. But his name wasn't *Vandenberg*. The Swedish prisoner was named *Wallenberg*.

Makinen repeated the names to himself like a mantra. *Vandenberg. Wallenberg. Vandenberg. Wallenberg.* It must have been lost in translation—Wallenberg had been pronounced *Vallenberg* so many times through the prison grapevine that it had slowly morphed into Vandenberg.

Makinen contacted one of the people mentioned in the article the very next morning. He called the front desk of the Linear Accelerator Center at Stanford University and asked for a high-energy physicist named Guy von Dardel, the son of Maj von Dardel and half brother of Raoul Wallenberg. He was transferred. He introduced himself. They hung up three hours later. From that day on Makinen would be part of the team searching for Wallenberg. He wouldn't stop looking for the rest of his life.

Makinen's realization coincided with a broader public acknowledgment of Wallenberg's bravery. By the late 1970s, Wallenberg's relatives had become demoralized. They had been investigating his disappearance for decades but were no closer to finding him, and Maj von Dardel's granddaughter once asked why she was still toiling away. "One can't accept the fact that a person just disappears," Wallenberg's mother said.

But one was becoming many. As the pressure on the Soviet government intensified, Wallenberg's living relatives were invited to Moscow by the KGB in 1989 to retrieve his possessions from the time of his arrest: his Lubyanka registration card, diplomatic passport, calendar, address book, cigarette case, and the money that he was carrying almost forty-five years earlier. An international committee organized by Guy von Dardel was later provided with documents of the Soviet prison system, the first time any group not affiliated with the government had been given such information. They were disappointed that Wallenberg was nowhere to be found in the prison records they inspected. But they weren't discouraged. They could sense the Russians offering a guise of cooperation during this period of glasnost. "To hinder the investigation of the case of Raoul Wallenberg is to stand on the wrong side of history," one Soviet minister told them.

The official party line was that Wallenberg, who exercised daily and had no family history of cardiac problems, had dropped dead of a heart attack when he was thirty-four years old. But now Makinen and his team of Wallenberg hunters finally had more than a false sense of hope of figuring out what the real story was. They also had a plan. They had been given permission to photograph the registration cards of roughly 900 prisoners: names, dates and places of birth, professions, nationalities, citizenships, criminal offenses, and the cells they occupied in the various buildings comprising the Vladimir Central Prison. These registration cards were the key to understanding how the prison worked. It made for the best data that had been collected in nearly a half century of searching for Wallenberg.

But they needed more data. They needed bigger data. They needed *better* data.

Makinen now believed that if their data was good enough, they could build a database of the Vladimir building where Wallenberg was reportedly imprisoned. Makinen argued to the Russians that he should be allowed to copy the registration cards of every prisoner who had spent as little as one day in this wing between the years of 1947 and 1972. This information might help them solve the mystery. "I knew in principle what could be done," he says. "I didn't really know how to do it."

He did not divulge to the Russians that he couldn't write the necessary software. He bluffed. He'd claimed that he could complete the analysis, and now he had to find someone who could. Someone who knew his way around computers. Someone who was fluent in the latest technology, versed in the scientific process, and comfortable with massive amounts of data. Someone like Ari Kaplan.

Ari Kaplan grew up in Lawrenceville, New Jersey, one of the American centers of the Jewish diaspora, where survivors of the Holocaust settled because of its proximity to New York City, to plentiful jobs, and to one another. Kaplan went to the California Institute of Technology, worked for Silicon Valley companies in the early internet era, and consulted for the Department of Defense as an intelligence contractor during the Gulf War. But his real specialty was sports. When he was a freshman at Caltech, Kaplan tried out for the baseball team. This being the Caltech baseball team, he soon found himself in the dugout wearing a uniform, an honest-to-goodness college athlete. He went to the plate four times. He struck out four times.

But what he lacked in baseball talent he made up for in baseball smarts. After his freshman year, before his tryout and long before any of his strikeouts, he did something that was more typical of a Caltech student: he applied for a research grant. The Summer Undergraduate Research Fellowship (SURF) Award was meant for the hard sciences, the type of stuff that appealed to Caltech undergrads, the people who would have rather passed their summers crunching numbers than riding waves. Kaplan spent his grant money studying Major League Baseball. The math that he was learning at Caltech dovetailed with the increasingly analytical thinking that was beginning to seep its way into professional sports, and he pored over box scores in the microfilm room of the nearest library to piece together a comprehensive history of relief pitching in baseball. The statistics that he unearthed were an effective way of thinking about a common problem—the problem in this case being the question of how to appropriately value the job of a relief pitcher in baseball.

This was heady research for the early 1990s. It would be more than a decade before a man named Bill James, the godfather of baseball sabermetrics, the process of applying statistical rigor to sports, was hired as a consultant to the Boston Red Sox and helped them win their first World Series in eighty-six years. Back then he was the kind of guy who wrote fan letters to professors like Amos Tversky. James was still at the point of his career when he was ostracized by people inside baseball, who felt threatened by the unknown, and lionized by people outside baseball, who were excited by the unknown. Kaplan fell in the latter category, and James was aware that he was becoming one of Kaplan's inspirations. "Our ability to generate stats has gotten way ahead of our ability to make any sense of it," he said at the time. "The first generation of computers gave us lots of numbers, but it's going to take . . . a lot of work by people like Mr. Kaplan before we understand what all this means."

But there *was* one power broker in baseball who liked Bill James, in part because this person was more familiar with the world of young Ari Kaplan. Eli Jacobs was a member of Caltech's board of trustees. When he attended the presentation of SURF Award winners, Jacobs paid closer attention than any of the other rich Caltech alumni in the room. He thought this kid who had studied the history of relief pitching in Major League Baseball could help with one of his recent investments. The man had just become the owner of the Baltimore Orioles. He hired Kaplan on the spot.

Kaplan's time with the Orioles taught him that it paid to hoard information. He worked with professional athletes who made more money than he did and grizzled executives who had more experience than he did. Information was his only form of currency. He took the filing cabinets where the Orioles stored their scouting reports and turned them into something with a fancier name than filing cabinet: a *database*. His computerization of information had a curious effect on the organization. The scouts who wrote their reports on computers began to get read. The scouts who scribbled in pencil began to get envious. They didn't take kindly to change, and they were suspicious if not outright contemptuous of the kid responsible for that change. They were all for certain kinds of information, but not his kind of information. Kaplan got used to being ignored.

But then Kaplan was introduced to someone who valued his input. His mother heard through a friend who had been Wallenberg's secretary in Budapest that Marvin Makinen was searching for a computer whiz. "My son would be perfect for this!" she said.

Kaplan was already familiar with Wallenberg's story when he visited Makinen's office, and the molecular biology professor briefed the twenty-five-year-old baseball dork on the status of his search for a missing war hero. Kaplan was eager to help, and Makinen needed all the help he could get. He'd recently had a breakthrough in his search for Wallenberg.

A warden in Vladimir had told him about an elderly woman who had worked in the prison since 1946, and she'd agreed to be interviewed in the place where she had spent every weekday since she was a teenager. Her name was Varvara Larina. She was cautious around Makinen at first. The staff hadn't been permitted to speak with outsiders about their jobs during the Soviet era. "We were not allowed to talk about the prison, and I will not do so now," a former guard had told Makinen on that same trip. But this interview was taking place in the office of the prison's chief physician. When she was hesitant, he said: "Tell what you know." Larina wasn't used to speaking about prisoners, especially with people she didn't know, and even more especially with explicit permission from her old bosses. But she relaxed once she was assured that she could say anything she wanted. It turned out that Varvara Larina had something remarkable to say.

Their interview in December 1993 began with a discussion of

her myriad jobs: cleaning prison cells, serving meager amounts of barely edible food, sterilizing equipment in the hospital wing. Larina didn't remember many of the people who had passed through the prison. That wasn't really her job. In fact it was specifically *not* her job. But when she was asked if she could recall any foreign prisoners, she admitted that she remembered one man. He was a prisoner kept in solitary confinement on the prison's third floor, and she recalled many things about him for someone who wasn't supposed to recall anything. He was a Westerner but not German. His arms were skinny. His fingers were long. His dark hair was thinning. And there was one more thing about him. He was in a cell opposite to one occupied by a man named Kirill Osmak.

Makinen asked why she remembered this prisoner out of all the prisoners.

"He complained constantly about everything," Larina said.

He was especially persnickety about soup. His soup was always cold because he was at the end of the food service. Even when it was hot, it was still prison soup. It was so thin he could see fish bones in the broth. ("And sometimes an eye," Makinen recalls from his own stint in Vladimir.) While this man complained constantly about everything, he really couldn't stand the cold soup. The prison's head guard finally snapped and ordered Larine to serve this prisoner first. She changed her entire route to accommodate one person. She had to ladle his soup first even if it meant climbing three flights of stairs with a heavy kettle before returning to feed the quieter prisoners on the first floor.

There were very few ways in which Ari Kaplan could relate to Varvara Larina. But when he was a kid, there was a woman on his paper route who woke at the crack of dawn, and she insisted that she get her *Times of Trenton* before anyone in the neighborhood. This was a colossal pain in Ari Kaplan's teenage butt. She lived at the end of the route, and Kaplan had to drag himself out of bed earlier than any boy would have liked. But he eventually yielded to her persistence. And he learned a lesson along the way. "You would remember somebody who complained day in and day out for months," he says. The frequency of her complaints left such an impression on Kaplan that he would never forget Ms. Kravitz from 18 Empress Lane, much like Varvara Larina would always remember this prisoner who forced her to change a daily routine.

It struck Makinen as unlikely that Vladimir's guards would ever grant any prisoner's wishes. He knew from his experience that ordinary prisoners would have been ignored if they complained about the temperature of soup. "Shut up! You're in prison" was a typical retort from the guards. A prisoner who complained as much as this man would have been sent to a punishment cell—bread and water every other day, and no sleeping blanket at night.

It was extraordinary for a prisoner to get his way. There must have been a reason that he was allowed to be different. Makinen took it upon himself to find out. He showed Larina a photographic lineup of fifteen men who looked nothing alike to see if any of them resembled the prisoner who hated cold soup.

"That's him!" she said.

She was pointing to a picture of Raoul Wallenberg.

It was a portrait from a side angle that only someone familiar with his appearance would have associated with him. That side angle was how Larina interacted with Wallenberg when he sat on his bed and she arrived with his soup of varying temperatures. It was a flabbergasting admission. But you wouldn't have known it by Makinen's reaction. He kept his cool and continued with the interview. It was even more important for him to maintain a poker face *because* an eyewitness like Larina had dropped a bombshell. The last thing he needed was for such valuable information to be tainted. Makinen didn't want Larina to believe she would be rewarded for saying anything that impressed him.

When he went back to Larina one year later to make sure she remembered Wallenberg the same way, he used the methods the Swedish diplomats had used on him after his return to the United States, a common trick in forensic investigations to assess how much a witness really knows. She told the same tale and picked the same photograph. That only made him want to come back *again*. Once again she told the same story. That she never embellished or changed any details made her testimony more credible in the eyes of Makinen. He even showed her digital re-creations from a forensic artist that depicted what Wallenberg might have looked like at different times of his life. She nodded at the one of Wallenberg in his midforties—the approximate age he would've been when his neighbor Osmak died.

What made this turn of events truly astonishing was that Osmak had died in May 1960. The Soviets were still claiming that Wallenberg died in 1947. There was no way around it. Someone was lying. Makinen was pretty sure it wasn't Larina.

Larina never knew the prisoner who demanded hot soup might have been named Raoul Wallenberg. She didn't even know who Raoul Wallenberg was. She couldn't have associated him with the righteous gentile who'd saved all those lives in the Holocaust. She had no reason not to tell the truth. If anything her incentive was to say the *exact opposite* of what she was saying.

The other piece of tantalizing evidence that startled Makinen came from a less reliable witness: a prisoner named Josif Terelya. Makinen interviewed him three times, once with a Ukrainian translator to make sure they fully understood each other, and it was important to fully understand Terelya, given his history of saying things only partially connected to reality. What he said was that one night in 1970 the Vladimir guard opened the door to his cell for Terelya and his cellmate to use the toilet at the end of the hall. The prisoners were supposed to visit the primitive toilet one cell at a time, but the guards were not always known for their attention to detail, especially not Terelya's favorite guard. His nickname was the Fool. "We called him that because he was flatulent and often passed gas loudly and with comical histrionics," he recalled. The Fool was on duty that night and opened Terelya's cell door seconds earlier than he should have.

That brief moment of indiscretion was all it took for Terelya to notice an elderly prisoner on his way back to cell 25. He'd never seen this prisoner before. Terelya could tell that he was a foreigner. He tried to keep track of his whereabouts from that point on, and he made a mental note to himself when the mysterious prisoner in cell 25 moved a week later to cell 33. The next time the Fool came around, Terelya asked him for a favor, the type of favor that only a prison guard named the Fool would have granted: he asked for the leftover cupboard in cell 25. The Fool obliged. Terelya searched the cupboard for clues of this prisoner's identity and found one written in purple ink. On the back of the cupboard were three words: "Raoul Wallenberg Sweden."

The only way for Marvin Makinen to verify this story was to study the data with Ari Kaplan. And to crack the mystery of Raoul Wallenberg they would have to go to prison. That's where the better data was.

In the case of Makinen, that meant going back to the same prison where he'd spent the worst years of his life. He'd felt tense with trauma the first time he returned to Russian soil in 1990. "After that I suppressed it," he says. "I had decided that I'm going to get this done. If I let myself be stressed by it, I wouldn't have gotten anything done." While it wasn't as stressful for Kaplan, he'd become intensely devoted to the search, too. "I was determined like nothing else to find out what happened to him," Kaplan says. "That's all I was focused on."

Every morning the two Americans had breakfast together in their hotel to discuss their goals for their grueling workdays. It was imperative that they had a plan for their allotted time in Moscow because there was only so much they could accomplish before they went back to their real jobs in the United States. It wasn't like they could take anything home with them. Their documents, their spreadsheets, their laptops—they were all stored in Moscow and couldn't be transported beyond Russian territory. Kaplan was in one room developing software and testing it. Makinen was in the other room examining the scanned registration cards against their computer representations. Everywhere they went, there were two guards. "One watching my left hand, and one watching my right hand," Kaplan says.

The Wallenberg detectives had negotiated access to a trove of data before they arrived. They spent a week in March 1998 scanning the brittle paper copies of registration cards for every Vladimir prisoner who had spent at least one day in the building where Wallenberg had been seen between 1947 and 1972—the period between his alleged death and when his possible sentence would have ended. That left them with 8,049 prisoners. Next they wanted to know which cells those 8,049 prisoners occupied. The registration cards had every prisoner's cell history—including when and where they were moved. This was crucial information. By the time Kaplan and Makinen and their team had transcribed the cards and transferred all that crucial information onto Hewlett-Packard laptops, they had a database of 98,030 cell records. They had created SportVU for Vladimir.

In their hands was a comprehensive history of a Soviet prison over twenty-five years. Marvin Makinen now had a proper understanding of who was in which cell on what day because Ari Kaplan had once again turned the contents of a filing cabinet into a computerized database.

But they didn't trust the data yet. The next step of their investigation was to check their work to make absolutely sure they hadn't come all the way to Russia only to make some truly unfortunate mistakes. They hired experts fluent in handwritten Russian who scrutinized the handwriting on the registration cards, and if they couldn't agree on a single character, they brought in one more expert to make a final call. Makinen and Kaplan took enormous pains to avoid human error. They considered their own biases and applied them to the minds of Vladimir's prison guards. If the records showed a prisoner moving cells on January 1, 1950, for example, they went out of their way to make sure a guard wasn't feeling the lingering effects of too much vodka from the night before. Did this prisoner *really* move on January 1, 1950? Or did it make more sense that he moved on January 1, 1951? Only by cross-referencing with other data could they resolve their questions. Such were the quality control measures Kaplan and Makinen implemented as they scrubbed the data.

"It was most important to us to be as objective as possible," Kaplan says. "We didn't want to skew the data. We wanted to be true to the data."

"So we just kept checking," Makinen says. "When you work in science, you carry out an experiment and get a result. It may be something that you expected. It may also be something that you didn't expect. But you still have to go back and check your technique. That's the way that Ari and I worked."

"The way you crack this," Kaplan says, *this* being a transcontinental conspiracy of monumental proportions, "is the unbiased scientific method."

They couldn't ask a dead Soviet prisoner if his cell history looked accurate. But they didn't have to. They asked Makinen instead. He remembered everything about his time in Vladimir. How could he forget? He didn't need to study the data to understand how frequently prisoners moved around (Makinen being Makinen, he studied it anyway) because he lived through it himself. He could relate to the prisoners whose names were now in a spreadsheet on Kaplan's computer. He could sympathize with their complaints. He could tell you the location of the medical clinic. He could remember the taste of the terrible soup. And he could explain *why* someone like Wallenberg complained. It had nothing to do with soup. It was about survival.

"He made sure that he stood out—that everyone was aware of him," Makinen says. "He was not allowing himself to be forgotten."

In that way he succeeded. Makinen and Kaplan devoted huge chunks of their lives to not allowing him to be forgotten. But not until they'd flown to Moscow several times a year for weeks at a time and spent thousands of hours on this wild-goose chase in which they didn't know if the wild goose actually existed did they begin looking for signals in the noise of their data. It was time to find out what happened to Raoul Wallenberg.

Among the eighty-two algorithms they wrote was one that would determine which prisoner was in the cell to which Larina served the hot soup. They had a feeling the answer would be fascinating. It was.

What they found was something they had always believed to be true. But until they'd done all that work, they could never know for sure. They didn't have the right data.

4.

On the night the NBA lockout ended, Brian Kopp's phone wouldn't stop ringing. Now that teams were finally allowed to make deals again, buying SportVU had become one of their top priorities. By the time he went to sleep, Kopp had deals with ten of the league's thirty teams to install tracking cameras in their arenas, and he could split the NBA into two groups. On one side were the teams with SportVU. On the other side were the teams without SportVU. On one side were the smart teams. On the other side were the stupid teams.

The divide became so glaring that teams on the stupid half started going out of their way to look smart. In the winter of 2013, the Los Angeles Lakers were the only team without an emissary at the Massachusetts Institute of Technology's Sloan Sports Analytics Conference, the annual excuse for NBA teams to brag about how smart they were, and they were instantly pegged as the stupidest of the stupid teams. It was a wonderful little twist. Now the nerds were bullying the kids who were too cool for school.

The Lakers decided they could no longer afford to look stupid after their public shaming, and they sought out Kopp to learn more about his SportVU cameras. Kopp noticed the general manager of the Lakers had a thick stack of papers on his desk. He couldn't help but peek at what he was reading: academic research. *Holy shit!* Kopp thought. He wasn't amazed that the top basketball decision-maker of the league's most glamorous franchise was reading academic research. What astonished him was the bizarre theatrical element of it all. Here was someone *performing* the reading of academic research. It was as if this NBA general manager had to be seen doing the work or else the whole exercise would've been worthless. This same general manager might have sprinted to the nearest paper shredder if he'd known the research on his desk was being produced by college kids between their classes.

When the Lakers were busy winning NBA championships, Carolyn Stein had been a student at a high school outside Boston. The only way into the honors science classes at her school was to participate in the annual science fair. It may have been the most highly competitive science fair on the planet. Stein's classmates were the children of biologists, chemists, and physicists accustomed to spending long hours in their parents' laboratories at Harvard and MIT. They were not making chemical volcanoes. "They were, like, sequencing genes," she says.

There were very few people at her school with such a sharp appreciation of their good fortune to come from a long line of scholars. Her grandfather was a Princeton mathematician who fled Europe in the 1940s and spent his first three weeks in the United States falling in love with the "strange game with sticks" better known as baseball, and her father was a Harvard economics professor who served on the U.S. Federal Reserve System while she was in college. But for this particular science fair, she had no special privileges. Her family couldn't help her sequence genes. Only after scrutinizing the rules of the science fair did Stein's father uncover a loophole buried deep in the fine print. The science fair technically allowed for projects in social sciences. "You're going to do a data project," he told her.

As the captain of her high school's basketball team, Stein decided to pour her energy into a data project about the NBA. "It wasn't that I loved sports," she says. "It was just that sports data was easy to get." She wrangled the numbers in her Microsoft Excel spreadsheets and regressed statistics like rebounds and blocks against a player's weight, race, and position. She called her project "White Men Can't Jump." "I almost got disqualified," Stein says. It was not like her to concoct a science-fair exhibit that managed to be controversial. "I was such a goody two-shoes kid," she says. "But I was just letting the numbers do the talking."

It was not exactly surprising that a high-school student producing intriguing, vaguely scandalous research would make her way to nearby Harvard, and one of the first things that Stein did after moving into her freshman dorm was attend the activities fair. She stumbled across an oddly named club: the Harvard Sports Analysis Collective (HSAC). She was more intrigued by analysis than sports. She went to a meeting, walked back to her dorm room, and told her neighbor John Ezekowitz that it seemed like a club that would appeal to him. Ezekowitz liked sports as much as analysis. A national Scrabble champion, he worked for the assistant Treasury secretary for economic policy when he was a teenager. He wasn't allowed to watch television at home when he was a child, but there was always one exception to this rule: sports. He watched a lot of sports.

The existence of this club at Harvard about sports analysis could be traced back to Michael Lewis's book about the Oakland Athletics and their reliance on data to search for inefficiencies in what should have been an efficient market. The influence of *Moneyball* is simply impossible to overstate. There's a line of demarcation in the history of professional sports: before *Moneyball* and after *Moneyball*. One person mentioned in the book was a Harvard statistics professor named Carl Morris who encouraged the students who came to him for advice to start a club about sports analytics. They didn't have to ask for a dean's permission to discuss all the ways in which their favorite sports teams were being transformed by data, but there was one benefit to registering as an officially sanctioned Harvard student organization: money. They spent that money on booze. "The club was founded," Ezekowitz says, "as an excuse to get free beer." That eventually changed. The college stopped funding the club, and the club stopped drinking beer.

The Harvard Sports Analysis Collective had been around for only a few years by the time Ezekowitz was elected president. It met in a stately room that appeared specially designed for afternoon tea, with its pastel wallpaper and many portraits of very old, very white men. Ezekowitz sat at the head of a long oak table surrounded by a bunch of bleary-eyed Harvard students. This was a meeting of dorks who liked talking about sports with other dorks. Ezekowitz called the meeting to order with a round of icebreakers. It was time for everyone in the room to introduce himself—or herself, but mostly himself—by naming a favorite sports book. The only catch was that it couldn't be *Moneyball*.

"How about the *Moneyball* screenplay?" someone cracked.

The youngest members of the club were about five years younger than I was when I visited in 2011. But what I realized when I looked around the oak table was that we belonged to different generations. In those five years, something had changed about the world, something that had never occurred to me before that night. *Moneyball* had been published. The book had been placed in their hands during their formative years as sports fans, which meant they were guided by a touchstone reassuring them that it was perfectly reasonable to make science of art. It was only natural that they would apply their smarts to sports. They belonged to the *Moneyball* generation. The idea that beliefs could and should be grounded in statistics didn't change the way these kids in the stuffy room at Harvard thought about sports. It *was* the way they thought about sports.

These college students didn't simply read *Moneyball*. They inhaled *Moneyball*. They lived and breathed *Moneyball*. And soon they embodied *Moneyball*. There was one passage in this book about finance, smack in the middle of a history of derivatives, that might as well have been their erotica. "The sort of people who quickly grasped the math of the matter were not typical traders," Lewis wrote. "They were highly trained mathematicians and statisticians and scientists who had abandoned whatever they were doing at Harvard or Stanford or MIT to make a killing on Wall Street. The fantastic sums of money hauled in by the sophisticated traders transformed the culture on Wall Street, and made quantitative analysis, as opposed to gut feel, the respectable way to go about making bets in the market."

The kids around the oak table could have been those ambitious traders, seduced by the opportunity to attain fabulous amounts of wealth, if not for the fact that swapping derivatives was no longer the most exciting opportunity available to them. They didn't have to exploit the inefficiencies of financial markets. They could apply their way of thinking to sports instead. It wouldn't be as insanely profitable, but it would be a whole lot more fun. These college students had the inclination, they certainly had the time, and, now, for the first time, they had the data. They could study anything that grabbed

their attention—even the ideas that had been litigated to death before they were alive. They could unsettle settled science. And they
would start with the hot hand.

It's not like some kids at Harvard were the first people to take issue
with certain parts of the original paper about the hot hand. Gilovich,
Vallone, and Tversky were bombarded with criticism from the second they published the study. That's partly *why* they published the
study. Tversky and Gilovich even wrote an explainer of their paper
in a 1989 issue of *Chance,* the official journal of the American Statistical Association. "The more basketball one watches," wrote two
people who watched a whole lot of basketball, "the more one encounters what appears to be streak shooting." But in a subsequent issue
of *Chance,* three statisticians wrote a biting response. "It's Okay to
Believe in the 'Hot Hand'" was the title of their paper. The evidence
they presented for the hot hand was so easy to dismiss that Tversky
and Gilovich were able to do it in the same issue of the journal. Pages
22 to 30 were one side of the argument. Pages 31 to 34 were the other
side.

The response that Tversky and Gilovich swatted aside was only
the first in a series of sustained attacks over several decades. The researchers in their wake weren't content with limiting their inquiry to
basketball. They reexamined the cases for and against the hot hand
in baseball, bowling, tennis, horseshoes, golf, volleyball, and darts. If
enough people believe that some kind of physical activity is a sport,
the odds are that it's been picked apart in search of the hot hand. The
most convincing studies did not question what Gilovich, Vallone,
and Tversky found so much as how they found it. The accusation that
stuck was that their samples were not large enough and their statistical tests were not powerful enough to detect the hot hand. Even if
there were such a thing as the hot hand, the paper that debunked the
phenomenon wouldn't have detected it.

Which brings us back to John Ezekowitz. By the end of his freshman year in college, he was poring through sports data and publishing ideas on the HSAC's blog that could help professional sports teams win games, and real people working for actual professional teams were reading him. Ezekowitz didn't know any of this until one Friday night in August, when he was neck-deep in a database that made his laptop crawl to a halt. His computer was acting like it had run a marathon on a summer day. It was hot, exhausted, and on the verge of passing out. When he published his blog post, it was a Friday night in August. He couldn't have picked a worse time to get eyeballs on the internet. For the next twenty-four hours almost nobody read it. But then Ezekowitz noticed a comment underneath the blog post. It was from Mark Cuban.

When he was done rubbing his eyes in disbelief, he decided that he must be doing something right if he was getting a response from an NBA owner. Perhaps this whole sports thing was worth pursuing. Maybe he could be a volunteer number cruncher for Harvard's basketball team. They couldn't pay him, and the data wouldn't be great, and his insights would probably be ignored. But still! It was a collection of athletes whose behavior might, in some way, at some point, be influenced by his brain. That was an intoxicating thought.

Ezekowitz was in his econometrics class one day when he received the phone call that would make all those plans seemed quaint. When his phone rang, he didn't recognize the area code. He decided to pick up anyway. He politely excused himself from class and stepped outside to answer. One of those real people from an actual professional sports team was on the line. The Phoenix Suns were calling.

The Suns were being pitched on SportVU by Brian Kopp, but they didn't know what to do with all the information they would soon be getting from the cameras in their arena. They used to have not enough data. Now they had too much data. The Suns had contacted a former president of HSAC who happened to be working for another sports team and begged him for a recommendation of someone who could help them sort through it. He gave them Eze-kowitz's name. The Suns didn't mind that the statistical consultant in their new analytics group was younger than their players, or that he was an undergraduate still taking Intro to Behavioral Economics, or that he worked remotely from an office thousands of miles away: his dorm room.

For the next few years, Ezekowitz spent his summer breaks around the Suns players, coaches, and executives, and eavesdrop-ping on their discussions was one of the great joys of his summer job. He felt obligated to pipe up one day as they debated the finer points of a subject that he thought he understood: the hot hand.

As someone who majored in economics and basically minored in basketball, Ezekowitz had read the classic Gilovich, Tversky, and Vallone paper. He'd been taught that people believed in the fallacy even after they were told it was a fallacy. Although he now worked with those people, he was still a little stunned about how strident they were. "There were tons of former NBA players and college play-ers in the room," he says. "They would all tell me to a man: *This is real.*" These were people he respected. They clearly knew more about basketball than he did. And it didn't offend Ezekowitz that so many of his colleagues disagreed with his intellectual heroes in such a fundamental way. It riveted him.

"There was this strange narrative of academics versus traditional guys, with the academics saying everything you think you know about the game isn't true and the traditional guys dismissing that," he says. "But the claims that academic guys make are fundamentally based on the data they have. As data becomes available, things we thought to be gospel may not be anymore."

He thought he could settle the debate if only he could find the right data. And then he realized that he already had it. Ezekowitz opened his laptop one afternoon during his winter break and typed out an email to his friend Carolyn Stein. "Hope exams are going/went well," he wrote. Ezekowitz and Stein were still living in the same dorm and taking the same economics classes. And he still owed her: she was the one who'd tipped him off to the Harvard Sports Analysis Collective in the first place. It was time for him to pay back his debt.

"I think previous hot-hand studies have all been flawed," he wrote. "The cool thing is that with the SportVU data I have, I think we can actually make a fairly definitive statement one way or the other. Do you have any interest in doing this as a class with me?"

5.

In his spare time, when he wasn't searching for Raoul Wallenberg or the molecules that identify cancer, Marvin Makinen was a heli-skier. He got his satisfaction gliding through powder and making fresh tracks. He didn't bother following trails that already existed. Instead he ripped down the side of a mountain and blasted his own path. He sought out the places he wasn't supposed to go. That's where the good stuff was.

The posters from ski trips of years past in his University of Chicago office were some of the only visible signs that Makinen was not like the other biochemistry and molecular biology professors in the building. You had to look beyond the shelves of dog-eared textbooks with names like *Classical Electrodynamics* and *Fourier Transform Infrared Spectrometry* and *Theoretical Foundations of Electron Spin Resonance* to notice the dusty box in the corner with photocopies of Soviet prison registration cards. Makinen was almost eighty years old, but he still had a sharp recall of every particular detail of his imprisonment. On the afternoon that I visit him with Ari Kaplan, Makinen reaches for a scrap of paper, a quiz about the essential amino acids. He flips the sheet over to its blank side and draws a map of Vladimir prison.

"Terelya was *here*," he says.

Josif Terelya's account was that the Fool opened cell 21 too early and allowed him to glimpse an older, Western, non-Russian prisoner walking back to cell 25 from the toilet at the end of the hall. This mysterious prisoner, the one who had written "Raoul Wallenberg Sweden" on his cupboard, was transferred to cell 33. Makinen had once been placed in that very cell. He had every reason not to believe Terelya, a mystic and self-proclaimed prophet, until he looked at the comprehensive database that Kaplan had built. Then he couldn't *not* believe Terelya. "What he told me was exactly, and I mean *precisely*, confirmed by the database," Makinen says.

Kaplan and Makinen had created a prison map for February 1, 1970. Terelya was in cell 21, and cell 25 was empty. But once cell 25 was occupied, cell 33 across the hall was empty. This indicated that the authorities had removed the documents identifying a prisoner. As if that evidence weren't convincing enough, this cell would be unoccupied for the next 117 days, and no other cell in this secluded wing of the prison went empty for more than five consecutive days. With a few keyboard taps on Makinen and Kaplan's antiquated laptops, Terelya's account had suddenly become credible. The data supported what he said. It showed that something fishy really was happening in cell 33. It was a statistical outlier.

The next person with a story worth checking was Varvara Larina's. She was the reason that Makinen had pressed for Vladimir's records to begin with. "This woman may have been mistaken," he says. "But we *had* to test this." It took much longer than Kaplan would have liked. The computers in prison drove him crazy, and basic queries that would have taken a few minutes to process on his computers back home in Chicago lasted twelve hours in Russia. He let his code go to work every night before he left and hoped that it would be done by the time they came back in the morning. One day Kaplan discovered his laptops were dead. The prison had lost electricity overnight.

It didn't take much forensic evidence to pinpoint the source of the outage. The electrical lines in the basement of the building had been chewed through. "We believe it was rats," he says.

But once the power was restored and they were confident the data was solid, Makinen and Kaplan clicked and waited for their slow computers to load a map of Vladimir on May 16, 1960. It was the day that Osmak died. Larina specifically remembered a prisoner who complained about cold soup living in one of two cells opposite his. If she misremembered, it would be obvious. There would be someone else there.

So who was in those cells?

"The cells were completely empty," Kaplan says.

The two cells across from Osmak when he died had been unoccupied for 243 or 274 days, at least if you believed Vladimir's records, which Makinen and Kaplan didn't. But just because they didn't believe them didn't necessarily mean they weren't believable. They went back to their records and checked how frequently cells were unoccupied for that long. In their twenty-five years of prison data, there were thousands of empty cells. But most of them were empty for brief periods of routine maintenance, and prisoners were back in those cells within a few days. The longest a cell was empty was seven days. The odds of a cell being unoccupied for 250 days was less than one trillionth of one percent. This was another statistical outlier.

But why would a cell be empty for 250 days? To answer their next question, Makinen and Kaplan singled out the prisoners whose profiles were similar to Wallenberg's. They could see that ordinary prisoners were generally transferred every 100 days, and the cells without identifiable occupants for 250 days or longer were extraordinary. One of the prisoners who moved cells every 200 to 300 days was the mayor of a city near the Polish border who had witnessed a terrible massacre the Soviets falsely blamed on the Germans. Because he couldn't be trusted, he was sentenced to twenty-five years in prison, the first seven of which he spent in solitary confinement. Makinen and Kaplan deduced from this pattern of movement that he was someone the Soviet regime took great lengths to isolate from the other prisoners. There was someone else who would have fit that profile: Raoul Wallenberg. This one anonymous prisoner who was transferred after 243 or 274 days was the closest thing to confirmation that Makinen and Kaplan would ever get. The odds of Larina being wrong about Wallenberg were infinitesimal.

It was a sublime moment. Makinen and Kaplan had turned false data into truth. Wallenberg had been expunged from the official records. They had found him anyway.

I have learned in my time interviewing scholars that it's silly to ask scientists if they have doubts about something. It doesn't even matter what that something is. Something can be anything. Of course the scientist has doubts. It's why scientists exist: to have doubts. The general rule of scientists is that the best ones have the most doubts.

But on this one topic, Makinen is unequivocal. There is no reason not to be. He'd collected the evidence. He'd analyzed the data. He felt comfortable making a conclusion. He might never know what really happened. But given what he did know, did he believe that Raoul Wallenberg was alive in prison for all those years he was supposed to be dead?

"I have no doubt," Marvin Makinen says.

6.

John Ezekowitz and Carolyn Stein were starting their independent study of the hot hand when a former president of Harvard walked into the basketball team's lounge one day to deliver an impromptu economics lecture. For the next thirty minutes, the basketball players listened to this professor in gym shorts and a gray White House shirt, a scholar once tapped to be the director of the National Economic Council and the secretary of the Treasury. As the Harvard players wolfed down pizza, Larry Summers reflected on his life in economics. There was a *New York Magazine* reporter in the room to capture the grand takeaway of Summers's career: "The key is reading data and recognizing what it tells you."

It was at that point in his unlikely seminar that a man whose name had been printed on money asked the Harvard basketball players how many of them believed in the hot hand. They all nodded. That was exactly the answer that Summers was expecting. He paused for dramatic effect before he revealed whether it was right to believe in the hot hand.

"The answer is no," he said. "People apply patterns to random data."

But not even the leading economic authority in the free world could have forecast that somewhere across campus a few undergraduate students were in the middle of figuring out whether that was actually true. They were reading the data and trying to recognize what it told them. They were attempting to figure out if the hot hand existed.

Almost everyone who hunted for the hot hand before Ezekowitz and Stein had acknowledged their analysis was flawed in a fundamental way. The studies that had found the hot hand to be a fallacy hadn't properly accounted for the fact that the chances of making a shot varied. "Each player has an ensemble of shots that vary in difficulty," Gilovich, Vallone, and Tversky had written, "and each shot is randomly selected from this ensemble." But was that really true? It seemed clear that player behavior changed when someone had the hot hand. The hot shooter was willing to take greater risks. His next shots were not necessarily coin flips. They could be dependent on the outcome of the previous shot. Remember what happened when Stephen Curry had the hot hand in Madison Square Garden.

For anyone to watch that performance and believe that every shot in basketball was the same was like believing you can dunk simply because you happen to have the same name as Michael Jordan. And yet the original investigations of the hot hand essentially hinged on that faulty logic. They didn't have a choice. They didn't have the data.

Only when Ezekowitz and Stein went to college were people able to study a topic like the hot hand with more sophistication. They could finally control for all those variables and accurately quantify the difficulty of a shot. That was the fantastically nerdy beauty of SportVU. If they could do it for one shot, they could do it for any number of shots. In fact they could do it for every shot taken in any NBA season.

They had discussed the intricacies of the hot hand when they took Intro to Behavioral Economics, and they had both come to question whether the canonical paper checked out all these years later. If a player thinks he's on fire, they agreed, he's going to push his limits and take harder shots. But the previous research hadn't taken this shift in behavior into account. The idea of doing something that nobody had been able to do before was alluring to Stein when Ezekowitz sent that email pitching her on the independent study. She wasn't the only one who wanted to work with him. There had been few people in the league who asked Brian Kopp sharper questions than this guy who was consulting for the Suns from his dorm room, and when Ezekowitz briefed him on his plans for a comprehensive study of the hot hand, Kopp barely hesitated before granting a bunch of Harvard undergraduates access to his database. "It was crazy that more academics didn't want it," Ezekowitz says, "because it was the richest data there was."

For hundreds of hours over the next semester, the students took it upon themselves to decipher that data. But first they had to clean it. While the SportVU data was rich, it was also messy. It did not lend itself to easy analysis. One of the other people who was given access to

SportVU's data happened to be a professor across the quad at Harvard. A formally trained cartographer, Kirk Goldsberry was floored when he opened his first SportVU file on his gigantic computer monitor. "All I could see was an ocean of decimal points, trailing digits, and hundreds of XML tags sporadically interleaved among them," he wrote. "Right away, it was obvious this was the 'biggest' data I had ever seen. I'll always remember my surprise when it occurred to me that everything on my screen amounted to only a few seconds of player action from one quarter of one game." As wonderful as it was, it was also worrisome to Ezekowitz and Stein. If even a fraction of the data was corrupted, their conclusions would be skewed. Marvin Makinen and Ari Kaplan knew the feeling.

By the time they were done, Ezekowitz, Stein, and their computer science expert, Andrew Bocskocsky, another Harvard undergraduate, had a better sense of the numbers that were changing the NBA than most NBA teams. This was because they had more data than Gilovich, Vallone, and Tversky could have imagined in their wildest, nerdiest, wonkiest dreams.

Once they plotted eighty-three thousand shots on a court divided into a grid of two-by-two squares, they could tell you anything you wanted to know about any shot from any NBA game. They could tell you where the shooter was, where the defender was, and where everyone else on the court was. They could tell you when the ball left the shooter's hand down to the twenty-fifth of a second. They could tell you how difficult the shot was. They could even tell you the probability of the shot falling through the hoop. And that meant they could tell you what they really wanted to say. "How is this possible?" Stein says. "For a long time, it really wasn't, short of watching hours of NBA footage and trying to guess how hard every shot was. You couldn't do it." Now you could.

The authors of the original, groundbreaking study of the hot hand had a small fraction of those shots. The quality of their data wasn't much better than the quantity. That paper treated layups the same as three-pointers, even if those shots have about as much in common as hippos and hamsters. It evaluated shots by hits and misses only. "Simple Heat" is what Ezekowitz, Stein, and Bocskocsky called that formula. But for a proper test of the hot hand, this "Simple Heat" wasn't powerful enough. It was a task that required something like "Complex Heat." That meant they would have to invent Complex Heat.

Since they knew everything about every shot—the identity of the shooter, the shooter's location, the defender's location, and the shot's difficulty—they could put a number on its probability. They called that number the "expected shooting percentage." Their calculation of expected shooting percentage was a useful conclusion in its own right, and NBA teams would soon begin using this metric that hadn't existed until they came up with it, which would have been a hugely satisfying outcome of nearly any other independent study that semester.

But for Ezekowitz and Stein, it was a means to an end. It was this number that was necessary for them to compute Complex Heat: the difference between *actual* shooting percentage and *expected* shooting percentage. Stephen Curry is obviously hotter when he makes five consecutive three-pointers than when he makes five straight layups, but only according to Complex Heat, not Simple Heat. Ezekowitz and Stein were the first to figure out how much hotter. Even when they had gotten that far, they understood how much further they had to go. "We had to spend time thinking about what questions we wanted to ask," Ezekowitz says.

This was not a trivial matter. Their answers would only be as good as their questions. They settled on two: Do basketball players change their behavior when someone appears to have a hot hand? And once you control for that change in behavior, does that hot hand appear?

The third question on their minds was one they couldn't quite articulate: What if everything they had been taught about the hot hand was wrong?

These were the sorts of questions that the forefathers of the *Moneyball* generation would have been asking. As it turned out, one of them was. Bill James had caught a glimpse of how teams made decisions in his time with the Red Sox, and it only made him more curious about the inner workings of sports. Right around the time Boston won the 2004 World Series, he published an essay titled "Underestimating the Fog" in the *Baseball Research Journal*. "If this was a real scientific journal and I was a real academic, the title of this article would be 'The Problem of Distinguishing Between Transient and Persistent Phenomena When Dealing with Variables from a Statistically Unstable Platform,'" he wrote. "But I was hoping somebody might actually read it." His wish came true, but not because of the title. It was because the paper itself was irresistible. "I have come to realize . . . that a wide range of conclusions in sabermetrics may be unfounded," wrote the godfather of sabermetrics.

His acolytes were not used to reading this sort of thing from James, who proceeded to question many of the fundamental truths of the statistical community that had all but elected him mayor. He wasn't saying they were untrue. He was saying that he couldn't be sure they were *actually* true. That someone of Bill James's stature would keep an open mind about the myths supposedly debunked by science might seem about as predictable as Betty Friedan proclaiming herself a misogynist. But in another way, it was quintessentially Bill James. The man who challenged the conventional wisdom of sports was now doubling back on himself and challenging the contrarian streak that was becoming the conventional wisdom. And you could almost hear him chortle as he typed the following passage about the most famous myth of them all.

"No one has made a compelling argument either in favor of or against the hot-hand phenomenon," James wrote. He continued:

> The hot-hand opponents are arguing—or seem to me to be arguing—that the absence of proof *is* proof. The absence of clear proof that hot hands exist is proof that they don't. I am arguing that it is not. The argument against hot streaks is based on the *assumption* that this analysis would detect hot streaks if they existed, rather than on the proven fact. Whether hot streaks exist or do not I do not know—but I think the assumption is false.

James then asked the types of people who pore through academic baseball journals in their spare time to imagine themselves on a battlefield. He wanted his readers to put themselves in the shoes of a watchman tasked with protecting his troops. You peer into the distance on a hazy night, but you don't see anything. You track down the brightest flashlight that you can point at the horizon. And still nothing. Only fog. You report back to your superior: *There is no enemy out there.*

But are you sure?

The absence of evidence is not evidence of absence. While the original hot-hand paper was interpreted as saying the hot hand didn't exist, what it really said was that *evidence* of the hot hand was absent. The authors had looked for the hot hand and seen the fog. "Let's look again," Bill James implored. "Let's give the fog a little more credit. Let's not be too sure that we haven't been missing something important."

Carolyn Stein and John Ezekowitz pointed the bright spotlight of their SportVU data into the foggy night. And suddenly it wasn't so foggy. They could see the rough contours of the hot hand coming into view.

They already knew the answer to their first question of whether basketball players altered their behavior when someone had the hot hand. The real question was if they could detect those changes in the data when they shined a light into the fog. And they could. They found that players took shots that were several feet farther away from the basket and that defenders crept a few inches closer when they felt the shooters were hot. Players were more likely to shoot when they had a hot hand, and the shots they took were more difficult. The assumption that the shots were randomly selected was a specious one. The shots were not independent. They were dependent on one another. This was the first proof that player behavior really did change when they felt the hot hand.

Once they established that much, Ezekowitz and Stein could turn their attention to the good stuff: Did that change in behavior mask the existence of a hot hand?

This is when their invention of Complex Heat came in handy. The hot hand was not about the number of shots that a player made in a row. Not entirely, anyway. The more reliable temperature reading of hotness was how much a player outperformed expectations. Their hypothesis from the second that Ezekowitz pressed send on his email to Stein pitching a study based on SportVU data was that a hot hand would emerge once they controlled for the difficulty of a shot.

Only at this point in their research, when they had painstakingly calculated that basketball shots were not randomly selected, did they have any confidence in the next part of their research: the shocking conclusion that there *was* such a thing as a hot hand. It amounted to a 1.2 percent improvement for players who made one of their past four shots and a 2.4 percent improvement when they made two of those shots—a small but nonetheless significant effect. That is, if a player made a few shots in a row, he wasn't less likely or the same amount of likely to make his next shot, at least not once you factored in the probability of making harder shots. He was slightly *more* likely. He was heating up. Then he was on fire.

While the result itself was modest, the meaning of it was monumental. Ezekowitz and Stein had found the best evidence of the famous hot-hand fallacy being a fallacy. Sometimes the truth was purposefully obfuscated by Soviet authorities. But sometimes the truth was simply obscured by data that wasn't good enough yet. Sometimes it was foggy.

"At the very least, our findings cast doubt on the overwhelming consensus that the hot hand is a fallacy," Bocskocsky, Ezekowitz, and Stein wrote. "Perhaps the next time a professor addresses the Harvard men's basketball team, the hot hand will not be so quickly dismissed."

On the day they presented their findings at the MIT Sloan Sports Analytics Conference, Stein received an email from someone who had plowed through their paper and dashed off a note from his iPad. He didn't need to introduce himself. "It's an impressive piece of work," Larry Summers wrote. "Congrats to you and your coauthors." He even took it upon himself to suggest lines of future research. "The broad issue here is variation in human performance," he wrote. "I feel smarter some days than others. Is this an illusion?"

The people who were the least impressed by this legitimately impressive research were Stein and Ezekowitz. Like any good scientists, they were skeptical about the findings, even if the findings were their own. They had doubts. Instead of overstating their results, they were purposefully cautious, politely reminding anyone who asked that they would've loved more data, better data, and that their results were conservative by design. The hot-hand effect they had uncovered was a small blaze and not the inferno of Mark Turmell's imagination. They were more comfortable defending their central finding that players really do change the way they play when someone on the court has the hot hand. They behave the way they believe. The undergraduates left it to a seasoned psychologist like Tom Gilovich to say that "this is by far the most interesting data I've seen supporting the idea that there is a hot hand" and a wizened economist like Larry Summers to say that "better data and better statistical techniques means we're going to understand the world much better." That people like Gilovich and Summers were discussing their work seriously was still bewildering to people like Stein and Ezekowitz.

"I don't think I'll ever do anything that other people care about as much as this," Stein says.

"I hope that's not true," Ezekowitz says.

"But it probably is," she says. "I don't know that in my academic career I'll ever have an idea that's so simple but people care about so much."

All they had done was exactly what they had been told to do. As it happens, it's what Marvin Makinen and Ari Kaplan had done, too. Carolyn Stein and John Ezekowitz had read the data and recognized what it told them. Their trick was that they had better data. And what their better data revealed was something it had never said before. It told them they might be right to believe in the hot hand.

———————

Seven

THE VAN GOGH IN THE ATTIC

"Aha, aha, aha!"

1.

It was a beautiful painting. In the beginning, long before Christian Mustad had any reason to believe there was a problem with this beautiful Vincent van Gogh painting of a sunset over a French abbey called Montmajour, he bought it because he liked the way it looked. And that's when his trouble began.

Mustad was the scion of a Norwegian company that manufactured everything from paper clips and zippers to fish hooks and horseshoe nails. And margarine. Mustad's factories also specialized in margarine. That was an important thing to know about him. The other thing that was important to know about this titan of industry was that he was a serious art collector. If you were a serious art collector in the twentieth century, you collected the works of Cézanne, Gauguin, Renoir, Degas, Munch, and, of course, Van Gogh.

The expert advising Mustad's purchases was Jens Thiis, the director of the National Gallery in Oslo, a trustworthy figure with a personal connection to this particular Van Gogh. Mustad needed someone like Thiis in his corner. There was an epidemic of forgeries sweeping Europe at the time, and jittery collectors would have believed anyone who sounded like he knew what he was talking about. But even art experts were right about the knockoff and the authentic Van Goghs as often as they were wrong. And sometimes they were wrong for reasons that had nothing to do with art. Mustad nevertheless had enough faith in Thiis's opinion that he decided to purchase this beautiful Van Gogh painting for his collection.

He would soon experience a nasty case of buyer's remorse.

The embarrassing ordeal that made him regret his prized acquisition began with a visit from his nemesis. Mustad had many reasons to be envious of Auguste Pellerin, his personal and professional rival, and chief among them was Pellerin's impressive art collection. He owned too many Cézannes to count. He lorded over a small mountain of Manets. He even had the odd Van Gogh. They both must have felt that Pellerin knew more about art than Mustad. As the Norwegian consul in Paris, Mustad ran in the same circles as the artists he supported. He was privy to their petty gossip, and he would have been familiar with the fears of fake Van Goghs.

But it wasn't merely Pellerin's art collection or his deep reservoir of art knowledge that had Mustad seething with envy. It was also the source of his fortune. Pellerin could afford his Cézannes, Manets, and the occasional Van Gogh because he made a killing as the owner of a manufacturing conglomerate. His company's name was Astra Margarine. Auguste Pellerin was another margarine tycoon.

Mustad's margarine enterprise was based in Norway and wanted to expand in France. Pellerin's was in France and wanted to expand in Norway. There was something delicious about two margarine titans being the owners of Van Gogh paintings during a time of incredible uncertainty about whether Van Gogh paintings were actually real. Mustad and Pellerin both had the money to buy Van Goghs from making something fake that seemed real. It wasn't butter. It was margarine.

That was more or less what Pellerin said as soon as he walked inside the home of his rival and Mustad showed off his proud new possession: the beautiful Van Gogh painting of a sunset over Montmajour. Pellerin told him that he'd been duped. One look at this painting was all it took for him to recognize an excellent forgery. It wasn't butter. It was margarine.

Mustad was mortified. He marched the painting to his attic, and that's where it would remain for the next half century. It didn't matter to Mustad that an expert he'd previously trusted had given this painting his stamp of approval. He didn't even bother asking for another opinion. The way that Pellerin dismissed the painting infuriated him to the point that he would never be able to look at it the same way again. The painting was forever spoiled.

Over the next fifty years, Mustad decorated the walls of his home with paintings by Munch, Cézanne, and Degas, turning his living room into a museum with a few sofas. He was even proud to display the other Van Goghs that he owned. But not the painting of a sunset over Montmajour. The art equivalent of margarine was hidden upstairs until the day that he died.

Christian Mustad was so humiliated about the whole episode that he went to his grave without having any clue that he may have been mistaken about the fake Van Gogh in one minor way. It wasn't fake.

2.

When he was a boy growing up in the suburbs of San Francisco, Josh Miller used to take the subway into the city. One day he visited the newsstands in Chinatown, bought some cheap explosives, and took the train home carrying a bagful of dynamite. He wanted to blow stuff up. Miller took his fireworks to the local basketball court, looked both ways to make sure no one else was around, and stuck one of his M-1000 fireworks inside a model car. He lit the fuse and ran as far as he could as fast as he could. When the detonation shattered his model car into millions of pieces, Miller was safely behind a tree on a hill, looking down on the damage. It was just about the coolest thing he'd ever seen.

Josh Miller loved fire. He was like Mark Turmell that way. His particular strain of pyromania took many forms. He roasted marshmallows on camping trips, fooled around with bottle rockets, and was only five years old when he accidentally set a tree ablaze. His intentions were always innocent. But what happened once he got started he could never control.

"You're just lighting stuff on fire," Miller says, "and then it gets bigger than you expect."

The kid who blew up a model car on the basketball court left his explosives at home when he went to college at the alma mater of Tom Gilovich. At the University of California at Santa Barbara, he met Adam Sanjurjo in Economics 101, a class that would be the spark for the intellectual exploration that would eventually make Miller and Sanjurjo the Lewis and Clark of the hot hand.

They had lots in common. They both came from Northern California. They both majored in economics and mathematics. They both went to graduate school, Miller to the University of Minnesota and Sanjurjo to the University of California at San Diego. But they both hit the academic job market immediately after the 2008 financial crash and smack in the depths of the recession. It was not exactly the best time to be an aspiring professor of economics. It may have been the worst time in nearly a century. Miller and Sanjurjo had to cross the ocean to find suitable work. Sanjurjo was half Spanish and spent his childhood summers in his father's native land. It had always been a goal of his to spend more time in Spain. A global fiscal crisis was a good excuse. He got a job at the University of Alicante and moved to the beach. Miller was in Milan and crashed in Sanjurjo's seaside apartment when he came to visit his college friend. So a

decade after their first class as undergraduates, Miller and Sanjurjo
had similar jobs in the same field. Sometimes they would head back
to California for a break and hang out in a cabin in scenic Marin
County owned by Miller's grandfather. They would wake up, have
breakfast, talk about ideas, work separately on their own projects,
meet for lunch, talk some more about ideas, work separately, make
dinner, drink wine, and talk even more about ideas. This routine was
so natural that it was almost as if there was a single circadian rhythm
bringing them together and begging them to collaborate.

"What the fuck are we doing?" Sanjurjo said at one point. "Why
aren't we working together?"

But first they had to figure out what they should work on together.
In the course of their many conversations, they came to realize they
both had been exposed to the research about the hot hand, and that's
when Miller and Sanjurjo let each other in on a dirty little secret:
they were both quite skeptical of the result. "I believe in the hot
hand," Sanjurjo wrote in an email to Miller in March 2010. He didn't
buy the argument that confidence had no effect on performance.
Miller agreed. It turned out that even their hunches were compat-
ible. Miller and Sanjurjo were as free as they would ever be. There
was no other place in the world that needed them and no other place
in the world that really wanted them. "It's time to transition from
talking about ideas," Miller said.

They decided to reopen the cold case of the hot hand. Sanjurjo remembered that he had a roundabout connection to a semipro Spanish basketball team that might be able to help. The team was more
semi- than pro. There are five rungs of basketball in Spain. The first
is the most competitive league in Europe with fantastic talents who
inevitably make their way to the NBA. But the quality of basketball
plunges from there. The players in the third division have day jobs
to pay the bills. There are intramural teams in the United States that
could beat semipro teams in Spain's fourth division. Miller and

Sanjurjo knew someone who knew someone who could help them run a shooting experiment with a team in the *fifth* division.

But the quality of players in their experiment was irrelevant. It was crucial for Miller and Sanjurjo to run a controlled test of shooting because they felt it was the ideal setting to detect the hot hand in basketball. There was too much noise that drowned out any signal in a real game. Only in an experiment could they eliminate the potential variables that dilute performance—the natural responses to catching fire that masked the appearance of the hot hand. Those were the things that Carolyn Stein and John Ezekowitz shined a light on: the difficulty of the shot, the quality of the player, the strategy of the defense, the score of the game as the crowd roared. There was no crowd in this case. Miller and Sanjurjo went to Betanzos, a small medieval Spanish city, and they had eight players take three hundred shots from a fixed position on the court where they shot about 50 percent. They ran the experiment again six months later with the same players.

This was similar to the shooting experiment in the Gilovich, Vallone, and Tversky study. But it was also different in a few critical ways. For one thing, there were more shots: three hundred per session instead of one hundred. They were also from the same location, not the same distance. And the players were shooting continuously instead of making bets before each shot. Miller and Sanjurjo also had data beyond this semipro Spanish basketball team. They looked at the shooters in the Gilovich, Vallone, and Tversky study and managed to track down a psychologist who kept the results of his own basketball experiments on computer punch cards. They hoped that a larger sample and greater statistical power would yield sharper results.

The previous shooting experiments made it difficult to measure the effect of the hot hand because they weren't equipped to detect subtlety. It was like trying to weigh coffee beans on a scale in a doctor's office. As they watched the videos of the Santo Domingo Betanzos players shooting in their empty arena, however, Miller and Sanjurjo felt they could tell when someone was hot or cold. But it didn't matter how they felt. The only thing that mattered to them was the math. And the math happened to agree with their feelings. The math showed that some players did have the hot hand.

"I refuse to believe this is self-delusion," Miller wrote Sanjurjo in an email.

"We are susceptible to cognitive illusion, but that only explains part of it," Sanjurjo responded. "There are also patterns. Maybe we blow the patterns that are there out of proportion. But there *are* patterns. This will be an improvement in the understanding of these processes over 'all pattern perception is cognitive illusion.'"

They worked on their paper over the next year and presented it to mixed reactions. Sanjurjo gave one talk at a conference in Toulouse, France, with Miller in the audience, where a Caltech behavioral economist was so respectful of their statistical chops and so engrossed by the specific details of their paper that he asked questions about a point they made in footnote 72. But they were ultimately disappointed in the response they got. One economist chided them in a gruff email that only a grizzled academic could have written. "In my experience, the strength of the results in a paper are inversely correlated with the amount of hyperbole used in the abstract," he wrote. "Using that heuristic, I am guessing that your results will not hold up. Next time, before you write an abstract, take a deep breath."

They exhaled and tried engaging him in polite discourse. That didn't work, either. This particular economist was not interested in hearing the nitty-gritty of their work. Whatever their shooting experiment with their sad collection of mediocre Spanish basketball players showed, there was almost no way that he was willing to change his mind about the hot hand. "If you ran an experiment in which some players did have a hot hand—so what?" he said. The gatekeepers who reviewed the paper for the most prestigious journals agreed with this person's assessment. "I just don't care very much if people can actually get hot when shooting or not," one wrote. Miller and Sanjurjo were told that the whole point of studying the hot hand was to show that humans see patterns where they don't exist. But even if they *did* exist, they still didn't exist to the extent that we believe they exist. "That is the real essence behind the hot-hand fallacy," this reviewer wrote.

As disappointed as they were, Miller and Sanjurjo refused to leave the hot hand in their rearview mirror. They needed to hear from one more person before they quit. Andrew Gelman was a Columbia University professor and the venerable statistician behind an eclectic, often downright eccentric, somehow wildly popular blog called *Statistical Modeling, Causal Inference, and Social Science*. As such he was intimately familiar with the history of the hot hand. Gelman was so fond of the original 1985 hot-hand paper that in his office filing cabinet he kept a preprint, a copy of the study before it cleared the peer review necessary for publication. "This paper came out and we immediately believed it: there is no hot hand, and people are wrong," Gelman says.

Maybe it was a surprise to people who didn't understand statistics. But not Gelman. When he taught the hot hand, he liked to split his classroom in two groups. The students in one group flipped a coin one hundred times and recorded the results—*H* for heads and *T* for tails. The students in the other group created a sequence that *looked* like they had flipped a coin one hundred times. Gelman would leave the classroom and come back to a blackboard that appeared something like this—let's say there were twenty flips instead of one hundred—and tell his students he could guess which was real and which was fake:

Group 1	Group 2
TTHHHTTTHHTTHTTTTHTT	THTTTHTHTTHHTTTHTHTT
TTHTTTHTTTHTTTTTTTTT	HHHTHTTHHTTTHHHHTHTT
TTHHTTHTHTHHTTTTHHHH	THHHTHTTTHHTTTHHTHTT

Gelman would study the sequences for a few seconds, pause for dramatic effect, and blow their minds. Group 1 was real. Group 2 was fake. This was the statistics professor's version of pulling a rabbit out of a hat. But how did he know?

"The real one is the one that looks fake," he says, "and the one that looks real is fake."

You've heard this before. The lesson that he was trying to impart was that random coin flips can appear streaky. That run of nine tails in a row in group 1? It's not how we imagine coin flips. But it still happens. Gelman scoffed at the people who insisted otherwise. "My take on all these people is they just can't handle reality," he says.

Miller gathered his courage and sent an email to his ideal reader with the paper he'd written with Sanjurjo that challenged Gelman's reality. "We have some new experimental and empirical work showing that the hot hand phenomenon can be substantial in individual players," he wrote. "Also, we find clear evidence of hot hand shooting in Gilovich, Vallone, and Tversky's original data set."

An obscure paper that disputed a classic finding of behavioral economics was the statistical blogger's equivalent of an adorable cat video. Gelman fired back a response to Miller and Sanjurjo on the same day. He thought it was clever work. It was not nothing. He admitted it might even be something. But a small, conservative, barely existent hot hand was not the *everything* they were after.

Who cares? Gelman thought.

Miller and Sanjurjo thought they had written a buzzy paper. They would've made a bigger splash cannonballing into the sea outside Sanjurjo's window. And nobody would have blamed them for pursuing another line of research and forgetting about the hot hand altogether. But it was still pulling at them. Their work felt incomplete for some unscientific reason that not even they could quantify. They were surprised by the pushback from economists willing to ignore the evidence they had presented. They believed more than ever before that the hot hand was real. They wanted to keep going. And so they did. They had been stung by the rejections from top economics journals, but as they went back to the drawing board, they remembered there had been some professors who had offered Miller and Sanjurjo a generous suggestion. They recommended looking for the hot hand in another kind of basketball game: the NBA's annual three-point shoot-out.

The league's best shooters assemble in one place every year for a shooting competition in which they have one minute to attempt five shots from five racks situated around the line. From the perspective of an economist, the three-point shoot-out wasn't simply a three-point shoot-out. It was a field experiment with tight controls and subjects who were the best shooters on earth—not the semipro players from Spain's fifth division. The three-point shoot-out was an exquisite testing site for the hot hand.

By now it had been almost two years since Miller and Sanjurjo made their first effort to study the hot hand. But good science takes a long time. It was about to take even longer. The only way they could make this study work to their satisfaction was to code every shot that had ever been taken in any NBA three-point shoot-out. They found most of them on YouTube and paid some guy in Switzerland for VHS tapes of the contests they couldn't find. They finished the arduous process of data collection only a few days after Stephen Curry (of course!) won the 2015 contest. When they were finally done, they had more than five thousand shots to sift through.

The first player whose performance Miller and Sanjurjo examined was a notoriously streaky shooter named Craig Hodges. Hodges's purpose in the NBA was to shoot three-pointers. Since this was long before there were many specialists like him, he was invited to participate in the contest for the first eight years of its existence. He won the 1990 shoot-out. He won the 1991 shoot-out and made nineteen shots in a row at one point. He won the 1992 shoot-out to complete his three-peat. Hodges was so integral to the three-point contest that the NBA let him into the 1993 shoot-out despite the fact that *he was no longer playing in the NBA*. It was that unimaginable to have one without him.

Hodges's legendary status in the three-point shoot-out meant that he was the most prolific contributor to Miller and Sanjurjo's database. They were so convinced that he would be on fire when they reviewed the footage that Sanjurjo offered to donate a critical part of his anatomy to charity if the numbers showed that he wasn't.

He wasn't.

Miller and Sanjurjo ran a quick analysis using tests from the Gilovich, Vallone, and Tversky study. According to this version of math, Hodges didn't have the hot hand. They went back to the footage, watched Hodges shoot again, and scratched their heads. It didn't make sense. The math clashed with the reality unfolding right in front of them. How could it be that Craig Hodges wasn't hot?

"There's something wrong with our brains," Sanjurjo said, "or there's something wrong with the statistic."

There was nothing wrong with their brains. The something wrong with the statistic had been hiding in plain sight the whole time. It was a subtle but crucial bias that had been there for anybody to discover. But nobody had. With the help of Craig Hodges, Josh Miller and Adam Sanjurjo looked at this particular set of facts, and they saw what everyone had missed. They had found the Van Gogh in the attic.

Now it was time to blow stuff up.

3.

On a cold night in the winter of 1888, Vincent van Gogh packed his bags and boarded an evening train in Paris bound for Arles, in the South of France. The fifteen months he spent there turned out to be the most concentrated stretch of creative success in his career. "The zenith, the climax, the greatest flowering of van Gogh's decade of artistic activity," one Van Gogh scholar would later write. Dashun Wang would call it something else: his hot-hand period.

Van Gogh's burst of inspiration would keep museums in business for many years to come. He produced more than three hundred paintings, watercolors, and sketches, masterpieces like *The Night Café* and *Starry Night Over the Rhône,* not to mention any of his other works now worth enough to buy a mansion in the South of France. His stint in Arles wasn't all sunflowers and starlight. He also went crazy and chopped off his ear. But it was in Arles where Vincent van Gogh became *Vincent van Gogh*.

That he'd picked Arles in the first place made about as much sense as a crocodile moving to a desert. He was looking for peace, quiet, and warm temperatures. Arles was freezing, windy, and snowy when he arrived near the end of February. There was nothing aesthetically inspiring about it. This environment unsuitable for making art left Van Gogh in a frustrating stalemate with himself, so he went for a hike one day to clear his mind. He settled on top of a hill and found himself admiring the panoramic view below him. It was a scene he wanted to paint. But he couldn't until the weather cooperated. He reminded himself to come back another time. "I've seen lots of beautiful things—a ruined abbey on a hill planted with hollies, pines and grey olive trees," he wrote to his brother, Theo. "We'll get down to that soon, I hope."

Then spring came. The weather turned. And summer arrived. One afternoon in early July, Van Gogh went for another sunset hike. He brought a canvas with him as he returned to the spot he'd described to Theo: a hill overlooking the ruins of Montmajour Abbey.

"I was on a stony heath where very small, twisted oaks grow, in the background a ruin on the hill, and wheatfields in the valley," he wrote to his brother the next day.

It was romantic, it couldn't be more so, à la Monticelli, the sun was pouring its very yellow rays over the bushes and the ground, absolutely a shower of gold. And all the lines were beautiful, the whole scene had a charming nobility. You wouldn't have been at all surprised to see knights and ladies suddenly appear, returning from hunting with hawks, or to hear the voice of an old Provençal troubadour. The fields seemed purple, the distances blue.

We can envision this view because Van Gogh didn't merely write it. He painted it. His blank canvas came alive with manic brushwork in creamy shades of white, green, blue, yellow, and red. But the next morning, when he sat down to write Theo, he examined the painting again. He hated it. "It was well below what I'd wished to do," he wrote. That the painting fell short of his ambitions said more about his ambitions than the painting. This would be one of the last times his aspirations were bigger than his imagination. Before long he would have paintings that made him proud—works of art that have endured for centuries. But not yet. Van Gogh was embarrassed by much of what he created before his breakthrough in Arles. He was so ashamed that he destroyed many of them. For some reason, however, he spared this one. About one month later, when he found himself firmly in the groove, Van Gogh sent a friend to Paris with a package of thirty-six oil paintings. "Among them are many with which I'm desperately dissatisfied," he wrote to Theo, "and which I'm sending you anyway because it will still give you a vague idea of some really fine subjects in the countryside." He would never again have to stare at this painting that disappointed him.

The painting with a view of Montmajour from Arles arrived safely in Paris, which is more than Van Gogh would be able to say for himself. In the aftermath of his death, this painting became indistinguishable from the others. It wasn't a substandard Van Gogh. It was simply a Van Gogh.

Now that he was dead, his family and friends were responsible for tracking the remaining, suddenly valuable Van Goghs, and the person who made the first attempt to catalog his inventory was the brother of Theo van Gogh's wife, a man named Andries Bonger. His effort is known as the Bonger List. He assigned a number to all 364 works that he could find. *Van Gogh's Chair* was 99. *Sunflowers* was 119. But the painting numbered 180 on the Bonger List wasn't as famous or instantly recognizable as those masterpieces. It was called *Soleil couchant à Arles*.

Sunset in Arles.

Bonger had no idea how useful it would be that he wrote 180 on the back of this canvas. By the time his sister loaned this painting to an exhibition in Amsterdam, it would not be known as *Sunset in Arles* for much longer. The painting's next stop was an artists' society that renamed it *Autumn Landscape* even though it was painted in the dead of summer. It soon vanished from the family's archives when it was sold to a French art dealer under the name *Groupe d'arbres avec nuages mouvementes,* or "Group of trees with scudding clouds." Van Gogh had been dead for almost two decades. His work was taking on a life of its own.

The painting was sold for the last time in 1908, and the next time anyone would see it was 1970. There was only one person who knew why it had disappeared: a Norwegian margarine titan named Christian Mustad.

Mustad wasn't in any state to explain the puzzling absence. He was dead, too. But real had become fake in all those years this Van Gogh painting had been languishing in Mustad's attic. The hike in Arles, the sunset on July 4, 1888, the painting's trip back to Paris, the creation of the Bonger List, the detours around Europe, the sale to a French art dealer—all of it had been wiped away from history. The facts had become fiction.

After his death in 1970, Mustad's heirs had his collection appraised, and one prominent art dealer agreed with Auguste Pellerin's assessment and told the family that his Van Gogh was a forgery. The painting wouldn't be evaluated again until 1991. That's when its owners sent the painting to people who would hopefully be able to say with certainty whether it was a Van Gogh: the researchers at the Van Gogh Museum. They dutifully inspected the canvas and offered the same verdict. "We think that the picture in question is *not* an authentic Van Gogh," they wrote. The fiction had become fact.

Auguste Pellerin had visited the home of Christian Mustad and scoffed at his purported Van Gogh in 1910. Now it was 2011. The painting in the attic had recently celebrated its centennial of inauthenticity. But in the two decades since it dismissed the painting, the Van Gogh Museum had matured. The first time it had been approached by the Mustad family, the museum was dealing with the most audacious theft of Van Goghs in modern times, a heist of twenty paintings that would be scarring for years to come, even if they were recovered less than an hour later. The experts were overwhelmed. They were too busy protecting real Van Goghs to spend time chasing Van Goghs that only might be real.

They were more amenable on the day the museum's senior researcher went digging around the archives. As someone who had devoted his career to studying one artist, Louis van Tilborgh could appreciate the glut of material that had become available to researchers in the time since Mustad's painting was last inspected. The scholars of his generation knew more about Van Gogh than anyone who came before them for two reasons. The first was that they had better data. The second was that they were willing to keep an open mind.

Van Tilborgh specialized in complicated questions of authenticity. Since it was his job to tell the difference between margarine and butter, it would have been malpractice if he *didn't* have an open mind. He had to be willing to litigate cases that had already been settled. He had to look at paintings that had been around for centuries and have the capacity to see them in a different light. He had to entertain the possibility that just because someone said something was real or fake didn't necessarily mean it was.

He was scouring the archives of the Van Gogh Museum when an old photograph of a painting happened to catch his eye. He'd been working at the museum for the better part of three decades, but he couldn't remember seeing this painting before. It was beautiful.

The officials at the Van Gogh Museum were not in the business of soliciting questionable Van Goghs. They would have been deluged with people who swore they had Van Goghs in their attics, and they were all too aware that it was rare to find a real Van Gogh that had been deemed fake. It wasn't worth their time. "We don't go after paintings," Van Tilborgh's colleague Teio Meedendorp says. "People come to us." But this one seemed like it had potential. Van Tilborgh decided that if anyone were to ever inquire about this beautiful Van

Gogh painting again, the least he could do was follow his intuition and take another look.

4.

Let's say Josh Miller is bored one day waiting for Adam Sanjurjo to meet him at the bar. His phone is dead. He pulls a coin from his pocket instead. He flips the coin.

Heads.

He flips the coin again.

Heads.

He flips the coin again.

Tails.

He keeps flipping the coin, and because he is a serious economist who knows better than to believe something that isn't backed by data, he grabs a napkin, asks the bartender for a pen, and jots down the results of every coin flip *after* he gets a head: *H* for heads and *T* for tails. Sanjurjo is so late that by the time he arrives Miller has flipped the coin one hundred times. They're intrigued. In this hypothetical universe, Sanjurjo pulls out his phone. They repeat their bar trick for ten thousand coins. They order their beers, read the napkin, check the phone, and study the *H*s—the heads after heads. They are shocked to see that their intuitive sense of randomness has led them astray. They realize that the proportion of heads after heads on a coin flip is not equal to the odds of getting heads on any old coin flip.

Miller and Sanjurjo weren't flipping coins or writing complex statistical formulas on bar napkins when they came across this particular mathematical bias. They weren't even in the same country. Miller was in Italy and Sanjurjo was in Spain as they watched old clips of Craig Hodges in the NBA's annual three-point shoot-out while speaking by Skype. The probability tests from past research suggested that Hodges wasn't hot. Their eyes suggested otherwise. "We *saw* Craig Hodges shoot," Sanjurjo says. "We know what we saw."

It was a necessarily defiant thought for a scientist. Maybe, just maybe, everyone else was wrong. Maybe, just maybe, they were right. This was almost exactly why the Gilovich, Vallone, and Tversky paper caused such pandemonium. That everyone could think they were right until they were suddenly proven wrong is also why the *New York Times* treated the findings of the original paper as news and included several pithy quotes from Amos Tversky in a write-up of the study.

"It may be that the only way you can learn about randomness," he concluded, "is to toss coins on the side while you play."

Miller and Sanjurjo took their intellectual ancestor at his word. They tossed coins. But you don't have to flip a coin hundreds of times to find what they found. You only have to do it three times. The short version of the math behind their breakthrough is simple enough to fit on a real bar napkin. Here is every possible outcome in a sequence of three coin flips:

TTT
TTH
THT
HTT
THH
HTH
HHT
HHH

Now let's take each of those three-flip sequences and look at the flips *after* heads flips. What percentage would you expect to be heads? It feels like the answer should be 50 percent—another coin flip. But let's average the results from the fifth column.

Sequence of Three Coin Flips	# of Flips After Heads	# of Heads on Those Flips	Heads After Heads	Percentage of Heads After Heads
TTT	0	0	-	-
TTH	0	0	-	-
THT	1	0	0/1	0%
HTT	1	0	0/1	0%
THH	1	1	1/1	100%
HTH	1	0	0/1	0%
HHT	2	1	1/2	50%
HHH	2	2	2/2	100%

The average is the sum of the fifth column divided by six. And what is 250 percent divided by six? It's not 50 percent. It's only 42 percent. Miller and Sanjurjo found the proportion of success after a streak is *less* than the underlying probability of success. If you were to generate a short, finite sequence like this string of coin flips and randomly select one of the heads, then the probability that the next flip would be a heads is closer to 40 percent than 50 percent. This is so trippy that Miller and Sanjuro could hardly believe it. Their brains weren't biased. The statistic was. They double-checked and triple-checked and would have quadruple-checked and quintuple-checked if they weren't already sure their careers were about to change forever.

"That was the most interesting intellectual moment of our relationship," Sanjurjo says, "and probably of my lifetime."

Only when they were done peeling their jaws off the floor could Miller and Sanjurjo wrap their heads around why this was such a big deal. I've watched them present this research in stuffy rooms on the campuses of Ivy League universities to academics who specialize in arcane fields. I've read the notes on their papers from peer reviewers empowered by academic journals to suss out the truth. I've followed the long and insufferable Twitter soliloquies from obnoxious people who can't wait to prove them wrong. The same thing tends to happen every time. No one believes them, at least not at first, and then everyone believes them. In that way, the Miller and Sanjurjo paper actually had a lot in common with the Gilovich, Vallone, and Tversky paper, even if it was asserting the opposite. "This infuriating and brilliant paper will prove that your intuitions about probability are not to be trusted," the economist Justin Wolfers once tweeted. That the paper was brilliant for the same reason it was infuriating—because it proved that intuitions about probability couldn't be trusted—was the best part of the whole thing.

The thoroughly original part of their discovery about coin flips went back to the original study of the hot hand. That paper had been assailed from the moment it was published, but most of the screaming and yelling about it was just that: noise. The big error was one that was undetected until Miller and Sanjurjo were confused by Craig Hodges.

The prevailing argument until they came along was that a basketball player's shooting percentage was unaffected by whether he was on a streak of successes or failures. That is, if a basketball player was a 50 percent shooter, and he was still a 50 percent shooter when he was hot, this was evidence against the hot hand. In fact that 50 percent was evidence *for* the hot hand. It's a subtle distinction and so barely perceptible that nobody had perceived it. "They've found something truly new—a serious mathematical flaw in the Gilovich-Tversky-Vallone study, missed by the many scientists, me included, who've combed through the paper in the 30 years since it came out," wrote Jordan Ellenberg, a renowned mathematician and the author of *How Not to Be Wrong*, in a piece for *Slate*.

The most impressive thing that Miller and Sanjurjo learned from their coin flips was that every other study of the hot hand had been statistically biased. The mindboggling truth was that 50 percent shooters should have been shooting *less* than 50 percent when they felt hot. Why? Because the proportion of hits after hits in a finite sequence is expected to be less than 50 percent. If a 50 percent shooter was shooting 50 percent, he was actually beating the odds. It didn't mean that he wasn't hot. It meant that he *was* hot. It meant the fallacy was a fallacy itself.

Once they accounted for this bias, Craig Hodges had the hot hand. Miller and Sanjurjo then reviewed the data from the seminal 1985 paper with their new statistical formulas, and those experiments showed that players were not less likely to make shots when they were hot. They were actually 12 percentage points *more* likely. That may not sound like much more than a dose of irony. What's 12 percentage points anyway? Well, let's put it this way: 12 percentage points is the difference between the average NBA shooter and Stephen Curry.

Miller and Sanjurjo were dumbstruck. This statistical bias they detected meant that the original study that reported no evidence of the hot hand had actually revealed hard evidence of the hot hand. They understood the consequences that came with overturning years of consensus. Any discovery of this magnitude would raise more questions than it answered. Why did nobody pick up on this before Miller and Sanjurjo? And what was anyone supposed to think about the hot hand now? "We don't have all the answers to all these questions," Sanjurjo says. "We're just trying to say everyone wasn't stupid. People were right to believe the hot hand exists."

They were still getting over their initial shock in April 2015, when Miller flew to Paris one afternoon, and Sanjurjo waited for him in a rental car. They were driving to a conference in Toulouse, France, and had given themselves a few days to make it there and enjoy the sights along the way. But they couldn't appreciate the beauty around them. They were too distracted. The only thing they wanted to do was talk about the hot hand. It was the only thing that mattered. In the same way that Amos Tversky and Daniel Kahneman once locked themselves in an office and became absorbed in their work, Josh Miller and Adam Sanjurjo briefly existed in this state of rapture where their oxygen was the idea. "Imagine being in the French countryside," Sanjurjo says, "but you're so excited about this idea that you're sitting in a room all day talking about it."

They managed to drag themselves away from their lodgings for long enough to hike, see a few castles, and scarf down escargot and foie gras. But otherwise they trapped themselves in their rooms to work. They were basking in the feeling of knowing a fact before anyone else in the world. "Is there any way we're wrong? No," Sanjurjo said. "Maybe I should be humble and say there's always the possibility we're wrong. But there's nothing wrong."

Yet having the facts on their side was no longer enough. In the three decades since the publication of the paper that started a field of hot-hand studies, the beliefs of most people had shifted. Now they had to follow in the same uphill footsteps as Gilovich, Vallone, and Tversky. They had to convince others they were right and pretty much everyone else was wrong. This wasn't their opinion. It was a mathematical proof. They knew the truth would always win over time and everyone would come to accept their work before long. But they couldn't wait that long.

From a hillside terrace in the medieval town of Autun, Miller and Sanjurjo plotted the unknown course ahead. The cathedrals below them had seen their fair share of conflict in the years since the Roman Empire. This remote Burgundy town was where caliphates ended and barbarians at the gates were turned away. And this was where two anonymous economists so far down the field's food chain that they didn't have jobs in their home country would hatch a plan to spread the word that everything that everyone thought about this classic study in behavioral economics was wrong.

"The battle is won," Miller said. "We just have to figure out how."

By the time they left Autun, they had a plan for how they would split the papers they had to write and who was going to work on what, and they knew it was going to work. They could see it all very clearly from the hills of France. The first thing they had to do was get in touch with someone back in the United States.

5.

It was still a beautiful painting. That much he could tell when the X-ray appeared on his computer.

Don Johnson wasn't an art historian. He was a professor of electrical and computer engineering at Rice University. He liked art and loved the Museum of Fine Arts down the street from his office in Houston, but he was by no means a trained expert like the specialists at the Van Gogh Museum in Amsterdam. And yet those experts had become his unlikely colleagues. Whenever they found themselves examining a questionable painting, they inevitably asked for Johnson's assistance.

He didn't know the history of Christian Mustad and Auguste Pellerin, and he certainly wasn't aware of the convoluted explanation for how an X-ray of this particular Van Gogh painting wound up on a computer in his engineering laboratory at Rice. "They didn't tell me a thing," he says. That was the point. He had to be able to look at the painting objectively and keep his work pure of preconception.

Johnson's work was counting paintings. To be more technically precise, he determined the pattern of thread separations in the canvas weave of paintings. He was an expert in the field of signal processing, and his opinion mattered to people who didn't even know what signal processing was: the art of extracting meaning from data. By feeding X-rays into the supercomputer on campus, he could determine which paintings came from which canvas rolls.

"It's not terribly useful for most artists, because the paintings have to be taken from the same canvas roll, and what artist buys canvas by the roll?" Johnson says. "Well, it turns out Van Gogh."

When he moved to Arles, Van Gogh worked on canvas rolls. He received those rolls from his brother in Paris and sent them back when there were enough for a proper shipment—even if he didn't particularly care for some of the paintings on those canvases. This turned out to be a highly useful habit if you were trying to prove the authenticity of a disputed piece of art. If you had to pick any work from any famous artist from any period of the nineteenth century, you would be wise to start with a Van Gogh painting from his creative peak in 1888. The new file on Johnson's computer happened to be one such painting.

He figured he would analyze the X-ray, compare the painting's thread count with the thread counts of roughly 450 paintings in his database, check if any of them were similar, and send the results back to the museum. The whole process of counting and comparing would take about a half hour. And he would probably hear nothing about this painting again. The novelty of seeing the guts of paintings from artistic geniuses was lost on him by now. Johnson could walk into any museum, gaze at the paintings on the wall, and consult his phone for a neatly organized spreadsheet listing every Van Gogh he's ever studied. "I've done so many of them that I can't remember," he says. He'd done it so many times that verifying a Van Gogh by applying the principles of engineering to art history had become routine. Don Johnson usually did his analysis without stopping to think about how cool it was.

But not this time. Because this time he got a hit. The X-ray of the unknown Van Gogh was an indirect match with a known Van Gogh painting called *The Rocks*.

Gee whiz, Johnson thought. *That's really weird.*

That it was *The Rocks* was not all that weird. *The Rocks* was another depiction of Montmajour from July 1888. Van Gogh's brother had pulled it from a package of paintings and framed it. The painting was given the Bonger List number 175. Similar scene, similar time, similar Bonger number to the painting on Johnson's computer.

There was nothing weird about that. The weird thing was that Johnson had seen *The Rocks* with his own eyes.

For as long as Johnson had been teaching at Rice, the crown jewel of Houston's Museum of Fine Arts had been a Van Gogh painting, a landscape from his 1888 summer in Arles. Johnson didn't have to open his handy spreadsheet to remember the name of that painting. It was called *The Rocks*.

The Van Gogh in the attic seemed to be from the same canvas roll as the Van Gogh down the street.

Johnson was so excited that he felt obligated to indicate some of this remarkable weirdness in the otherwise dry analysis he sent back to the Van Gogh Museum.

June 19, 2012

I have counted E1657. The report will emerge soon; a little backed up at the moment. But here's the news.

No weave match. BUT, the canvas has all the appearance of coming from a "typical" Arles pre-primed roll. I see strong cusping on one side and a warp thread repair . . . I found a count match with another Arles painting (F466) that also has no weave matches. (A count match means that the canvases are VERY similar but does not mean that two count-match paintings had to come from the same roll/bolt.)

This was a fairly standard technical report so far. But there was one exclamatory line in his email that Don Johnson never had the pleasure of writing in any of his previous reports.

Turns out that F466 is located in Houston!!

Once he pressed send on that email, Don Johnson went on with his day, and it was a good thing that he wasn't anxious to hear back from the Van Gogh Museum. It would be quite some time before he got word of what researchers like Teio Meedendorp did with his work.

Meedendorp was unaware of the unknown Van Gogh painting's existence until the series of events that would culminate in the high-light of his professional life. It had started when Louis van Tilborgh stumbled upon a photograph in the archives of the museum. What happened next is still protected by the shroud of secrecy that tends to follow paintings from centuries ago that are now worth millions of dollars. The only thing museum officials will say is that the call that they were awaiting came from a friend of the painting's owner who happened to be from Arles and recognized the view in this painting and happened to be familiar with Van Gogh's letters and the descrip-tion of his sunset hike on July 4, 1888. It was time for the Van Gogh Museum's researchers to take a more serious look at this painting.

The first thing they looked at was the painting itself. Once they gave it the benefit of the doubt, they began to notice things they'd previously missed. They were able to confirm through technical research that the paint matched the oils that Van Gogh used in Arles. The colors were the shades of his palette that summer, and the brushwork was sloppy because of the wind. Those clues gave them an incentive to look for more. They took a trip to Arles in search of the sunset view and reread the letters, nearly one thousand in total, that Van Gogh researchers had stared at for so long they could recite them by heart. The museum's online collection of fully annotated letters made it easier than ever to connect the dots, and when they went back to the letters, there it was in his note to his brother on July 5, 1888: Van Gogh's description of this painting.

Why hadn't they seen it earlier? They had. They just hadn't known what they were looking at. It was right in front of their eyes, but even the experts had missed it.

Van Gogh's description of the painting had always been attributed to *another* painting. Only in retrospect did they realize how phenomenally wrong they were. The abbey ruins were missing from this other painting. So were the "wheatfields in the valley," the "small, twisted oaks," and the "very yellow rays over the bushes." The scholars had always seen what they wanted to see even if it wasn't actually there. The painting they believed to be the one in Van Gogh's letter would have looked familiar to Don Johnson. It was *The Rocks*—the one in Houston made from the same roll of canvas. They thought the Van Gogh down the street was the Van Gogh in the attic. Now maybe it wasn't.

The next box on their checklist brought the researchers back to the Bonger List. By now there weren't many unidentified paintings on Andries Bonger's comprehensive accounting of Van Gogh's work from a century earlier. Only a few of the 364 were missing. The most suspicious of the missing works from the Bonger List was number 180.

"Is there anything on the back?" Meedendorp asked Van Tilborgh.

They looked at the original technical report for this painting. Indeed there was something on the back. And it was something they would have to see to believe for themselves. They turned the painting around to find three numerals in permanent ink: *180*.

To confirm they hadn't gone temporarily crazy, they took another glance at the Bonger List, and there it was again: "180 (*Soleil couchant à Arles*)." You didn't have to be one of the world's leading Van Gogh scholars to realize this painting with the number 180 on the back sure looked like a sunset in Arles.

"It adds up and it adds up and it adds up," Meedendorp says. "It's like going, *Aha, aha, aha!*"

To believe something extraordinary demands an extraordinary burden of proof. But the small pile of evidence in favor of this painting being a real Van Gogh quickly became difficult to ignore. "It is still incomprehensible how what seems to have been the first picture by Van Gogh to enter a private Norwegian collection managed to escape mention in the literature on Van Gogh for so long," Meedendorp, Van Tilborgh, and Oda van Maanen would later write.

But was it? Maybe it was more comprehensible than they acknowledged. For too long people had been looking at this Van Gogh the wrong way. Auguste Pellerin was driven by envy when he informed his rival that his painting was a forgery. Christian Mustad was deluded by his own self-doubt when he marched the Van Gogh to his attic. The art dealers, art connoisseurs, and art historians believed what they had always been told. But what if they had started from scratch? There would have been no prior assumptions for them to overturn. The experts would have looked at this painting and seen the truth staring them in the face the whole time.

Don Johnson woke up on a Monday in September 2013, more than a year after sharing his analysis of a painting that seemed to be a cousin of *The Rocks* with the Van Gogh Museum, and he rolled out of bed to check his inbox. As he scrolled through the spam of a Monday morning, his attention gravitated toward an email that had been sent at 3:55 A.M. local time. And suddenly he wasn't sleepy. "New discovery Van Gogh" was the subject line of the email that jolted him awake. His friends at the museum wanted him to know they were about to unveil a new painting called *Sunset at Montmajour*.

The Van Gogh in the attic had been real. Then it was fake. Now it was real again.

6.

One morning in July 2015, a few months after his trip to the French town of Autun with his intellectual coconspirator Adam Sanjurjo, Josh Miller sent an important email of his own.

He was on his way to New York to give a presentation to Microsoft's research division when he thought of a local celebrity who might be interested in his latest thoughts about the hot hand. He got in touch with Andrew Gelman again. Gelman is someone who doesn't get upset when it appears the entire world might be wrong about something. He gets curious. He wants to know why it's wrong and how it came to be wrong. "When we see holes in our theories, or contradictions, or anomalies, we should be bothered by these things rather than trying to explain them away," he says. That capacity for discomfort is the mark of an honest scientist. Gelman believes that people in his profession have to seek out challenges to their own beliefs. They have to be willing to change their minds.

Gelman read Miller and Sanjurjo's paper, digested the statistical bias that even this statistician had missed, and abided by his professional standards: he changed his mind about the hot hand. "This is pretty obvious once you think about it," he says. "But I hadn't thought about it."

There is great pleasure to be found in the investigation of potential wrongness. When he invited Miller to his office the next day, Gelman typed out simple lines of code on his computer just to be sure. Miller grinned. "No one believes me," he said. But the computer did. The computer had no skin in the game and only cared about the math being right. Miller and Sanjurjo knew by now that it was. That was the confirmation Gelman required to bring him back to his office the next day. He arrived as Gelman was putting the finishing touches on the blog post announcing his support for Miller and Sanjurjo's wildly contrarian study.

They were years away from officially publishing in the most prestigious journal of economic theory that once published Amos Tversky and Daniel Kahneman. All they had so far was a working paper uploaded to a free website that served as a salt flat of ideas: a place for academics to road test their latest thoughts and accelerate the leisurely publishing cycle to speeds that were once inconceivable. Instead of waiting years for a panel of distinguished scholars to referee their paper and rubber-stamp the math that Gelman's computer had proven to be accurate, now anyone could read Miller and Sanjurjo's calculations for themselves. Anyone turned out to be everyone when Gelman hit publish. In a blog post titled "Hey— Guess What? There Really Is a Hot Hand," he wrote, "No, it's not April 1, and yup, I'm serious." Andrew Gelman was their peer review.

What happened next was a crash course in the way the internet works. As soon as Gelman wrote about Miller and Sanjurjo, all hell broke loose on his blog. The rules of decorum that governed conversations about math went ignored by thousands of enraged psychologists, economists, and statisticians. Miller and Sanjurjo spent hours playing whack-a-mole in the comments section of Gelman's blog. Every time they responded to one hostile challenge to their paper, another string of complaints appeared. They would've gotten a friendlier reaction if they'd written a technically impenetrable paper arguing that puppies are moral abominations. Gelman got a kick out of the comments section becoming a war zone. "It's just math," he says. "It freaks people out."

The only problem with the spectacular popularity of their paper was that Miller and Sanjurjo still hadn't published it anywhere. That meant it was still illegitimate in the eyes of their fellow academics who allowed people like them into the firmament of economic thought. And *that* meant Miller and Sanjurjo would have to bring their paper on the road and convince people who weren't Andrew Gelman. They presented the saga of the hot hand over samosas at Yale University ("the hot hand exists, is meaningfully large, and is likely underestimated") and finger sandwiches at New York University ("the hot hand is neither a myth nor a cognitive illusion"), and it was during this world tour that a semiretired statistician named Bob

Wardrop noticed an advertisement for Miller's visit to the University of Wisconsin campus.

Wardrop had objected to the original hot-hand work in the 1990s by publishing two papers questioning the statistical methods of the study. They hadn't fallen on deaf ears. Like everyone else who dabbled in the hot hand, Wardrop says those papers were his most contentious. But they hadn't inspired a civil war among scholars, either. Wardrop was more offended than most by the idea that the hot hand was a fallacy. It fit into his grand unified theory of how academics viewed their place in the world. "Everyone is dumb, and we should poke fun at them," he says. "One of the ways you can always make a living as an academic is to write papers about how stupid regular people are." It was his first thought when he read those papers that claimed the hot hand didn't exist. Wardrop came from blue-collar roots and paid his way through college working in Detroit's car factories, which shaped the way that he thought once he made thinking his day job. "You have to understand that you might have a different perspective than the people you're questioning," he says. "You might have different data."

As a result of his own experiences, Wardrop came down on the side that professional practitioners had an expertise that could not be overstated or fully understood. He didn't think they were as stupid as smart people thought they were. Wardrop had long since made peace with his place in the history of the hot hand when he attended Miller's talk at Wisconsin. He listened as Miller carefully explained the statistical bias and argued that belief in the hot hand is justified. "He just blew the lid off a really major mistake," Wardrop says, "which embarrassingly I should've noticed." Wardrop had noted the hot hand was an occasional phenomenon that was too often treated as an omnipotent phenomenon. But this bias had always been there, and even a statistician who had been skeptical had somehow missed it. "They were able to see what the problem was,"

Wardrop says. "Even though my heart was pure and my intentions were good, I could not."

Miller came back to Columbia at the end of his tour. It was a gray spring day inside a dreary lecture hall. The school year was almost over. There wasn't much advertising for the talk, which may explain why the room was almost completely empty. He was speaking to a crowd of no more than a dozen people.

But it didn't matter how many of them there were because of *who* they were. In this room to hear about Miller and Sanjurjo's study of the hot hand was a dazzling array of intellectuals. In the far corner of the room was Andrew Gelman. He could have charged Miller and Sanjurjo for all the publicity that he'd provided them. A few rows behind Gelman was the bestselling author Nassim Nicholas Taleb. He was familiar with the hot hand from his books about randomness and probability. In the dead center of the front row was an older white-haired man with the most impeccable credentials of them all. In his neat suit and black trench coat, he was striking. Even if you didn't know who he was, you would have guessed that he was someone with gravitas, the type of person whose presence in this shabby room meant something. Josh Miller did know who he was. He was about to deliver his lecture directly to Daniel Kahneman.

Miller didn't appear nervous even as his insides were doing the Electric Slide. He offered a coherent definition of the hot hand and traced the history of his subject as he explained the bias to his rapt audience. Kahneman was locked in the whole time. Over an hour later, right before the round of applause, Miller concluded with a thought that everyone in this lecture hall on a weekday morning could agree with. "There's always going to be a certain mystery to it," he said.

There were a few brave souls who raised their hands to ask questions, but the canny ones knew better than to say anything yet. Before a verdict could be reached in what had turned into a trial of the

hot hand, the jurors wanted to hear from Kahneman. By this point
in his career, Kahneman was the kind of guy that Tom Gilovich
would have begged for a selfie on his first day of graduate school.
He'd taken the ideas that he'd formulated with Amos Tversky and
translated their papers that had won the Nobel Prize for a general
audience. *Thinking, Fast and Slow,* the culmination of his life's work
in book form, was a huge bestseller. The result was that his sheer
presence in the room had a chilling effect. No one was willing to
admit they believed something that Kahneman didn't.

Kahneman hadn't just stumbled into a random classroom to hear
this talk about the hot hand. He was aware of Miller and Sanjurjo's
work. He knew that Israeli statistician Yosef Rinott and psychologist
Maya Bar-Hillel had published a comment about their paper that
supported their conclusion, confirming that Miller and Sanjurjo
had "raised a valid and overlooked criticism," one that was enough
to "promise that the hot-hand debate will continue." Now that was
about to happen. The air rushed out of the room when he raised his
hand.

"A couple of questions," he said.

Kahneman started by defending Tversky's immense legacy
and stressing the underlying principle that when you show people
randomness, they don't believe it's random. Miller and Sanjurjo
had heard this plenty of times before. But what he said next was
something astonishing they had been waiting for years to hear.

"I think clearly Tversky et al. were wrong," Kahneman said. "Their
test was biased, and there is a hot hand.

"It was an unfortunate thing they went on to make that mistake,"
he continued. "But the point is still valid. People see patterns where
there are none."

In this one brief exchange was the entire story of the hot hand.
Every good idea should invite criticism. This one was now being
questioned by the statistical insights of economists who owed a debt

to psychologists. Miller and Sanjurjo had breathed new life into an idea by looking at it through the lenses of all three disciplines.

The few people in the room that day weren't the only ones who would change their minds about the hot hand. Miller and Sanjurjo proved to be excellent at making skeptics come around to their perspective, and their arguments were so persuasive that it wasn't long before the editors of a top journal of knotty economic ideas agreed with them. In the November 2018 issue of *Econometrica,* finally, was a paper called "Surprised by the Hot Hand Fallacy? A Truth in the Law of Small Numbers," by Joshua B. Miller and Adam Sanjurjo.

If so many people could be mistaken for so long about this, then what else have we been wrong about? Of course there are times when experts are blinded by their own expertise. But there are other times when they might actually know what they're doing. Those are the times when it might be worth listening to Jens Thiis when he says the Van Gogh is real and Stephen Curry when he insists that he has the hot hand.

The opportunity to see this physical manifestation of his work in person was one of the many reasons that I took Miller to a Golden State Warriors game when we both happened to be in the Bay Area one evening in April 2016. It was a fortuitous time to be there. The Warriors were shattering NBA records every night on their way to becoming one of the greatest basketball team ever. They were a week away from winning more games than any team in the history of the league. Why? It was because they had empowered Stephen Curry to keep shooting after the night he caught fire. I joined Miller in one of the last rows of the upper deck of the arena, so far away from the court that we could touch the roof of the building, as the game somewhere below us started.

It was not Curry's finest hour. He missed his first eight shots of the night. He missed layups. He missed three-pointers. We had managed to pick the one Warriors game in which Curry had the worst half in maybe the best individual season of all time. The greatest shooter ever couldn't make a shot.

But early in the third quarter, at the exact moment it seemed like Miller had a better chance of hitting a three-pointer from our seats in the heavens, Curry seemed to forget that he was supposed to be frozen cold. He fired a three-pointer. This one he made—his first shot of the night. He strutted back to his bench with his arms extended and his palms upward, as if to say, *It's about time.* The other team had seen this before and immediately called a time-out. The last thing it needed was for Curry to get the confidence to make another one. It was too late.

He made another shot with a giant human stampeding toward him. He made another shot over the hand of a defender so close to his face that he couldn't see the basket. He made one more shot after dancing away from the defense and creating the teeniest sliver of an opening to launch the kind of shot that nobody else who had ever played basketball would have even bothered attempting. The ball was still in the air when he started moonwalking away from the basket. He could see the future. He already knew that his shot was going in. Of course it was. Stephen Curry had the hot hand.

Josh Miller cackled into his beer so hard that it nearly tipped over. "Look at that!" he yelled. He was staring at something that wasn't supposed to exist. It was a marvelous sight.

EPILOGUE

"Tom Gilovich," said Tom Gilovich.

"Matt," said the anonymous research subject.

"Nice to meet you!"

It was a gray autumn day on the campus of Cornell University when I walked into Gilovich's spacious office. On the massive bookshelf behind his desk, beneath the social psychology textbooks and between his own books about happiness, there was a basketball that looked like it could have been handled by James Naismith himself. It wasn't quite that old. This basketball on his bookshelf was a souvenir. It was the one that Gilovich had used in his 1980s experiments about the hot hand. It had been more than three decades since the publication of his breakthrough paper, and so much had changed about the public's thinking about the hot hand that Gilovich had been feeling inspired lately. That's why I had come to Cornell. I was here to watch him run another shooting experiment.

He stomped through leaves on this first cold day of fall as he made his way to the Cornell gym. This was the type of weather that took some adjusting to when he left the West Coast and never went back. There were some things that he missed about California in his early years on the East Coast. The vegetables, for example. The vegetables were a big problem. When he came to upstate New York, where it can feel like winter in fall and spring, Gilovich was disappointed in the local produce. But things change. "Now we have a Wegmans," he says. And it was because of things changing that Gilovich was hosting me that afternoon.

He'd been at Cornell long enough that he'd developed a reputation as a legend in his own right. Now he was the same age as the professors at Stanford he'd once idolized, and Gilovich had more salt than pepper in his hair to show for it. He was also as congenial as his mentors had been. Gilovich was the sort of person who signed his emails "Cheers" and seemed like he might have stopped typing for a second to raise his cup of coffee. The other thing that he had in common with his Stanford professors was that he was quite familiar with the basketball court near his office. He'd spent many hours in Cornell's gym playing in lunchtime games. Only recently had he stopped. Ever since he retired from basketball, he kept himself busy watching his beloved Boston Celtics, the same team that Red Auerbach was coaching when he questioned Gilovich's famous paper with Robert Vallone and Amos Tversky.

It shouldn't have been a surprise that someone who'd made a career of studying human judgment and decision-making had a theory about pickup basketball. Gilovich believed the best games were the ones in which the best players were the best passers. "Cornell had the best noontime game for twenty-five years," he says. But the players who made the game so great were long gone when Gilovich stopped playing. They got hurt. They moved away. They were replaced by players who didn't care about passing as much. Things changed. "The last couple years that I played, it was still a good game, but it didn't have the magic of that great run," he says. Gilovich was on the wrong end of his rule by the end of his playing days. "In general," he says, "people aren't going to pass the ball to a gray-haired guy."

But the nice thing about being the gray-haired guy in a gym is that you have the respect of everyone around you. When he walked into this gym, Gilovich found his undergraduate research assistant

Willie waiting for him. Max and Matt, two Cornell basketball players, apologized profusely when they arrived a few minutes later. They
came in gym shorts and sneakers even though they didn't exactly
know what the experiment entailed, and they stood underneath the
basket as Willie explained they were about to be participants in this
argument that had been raging since long before they were born.
Willie asked them to pick seven locations on the court where they
felt they were 50 percent shooters with nobody guarding them. Matt
and Max both picked spots dispersed around the three-point line—
the area of the court that Stephen Curry had colonized.

It was amazing the degree to which history could be rewritten
over the course of one generation. David Booth was molded by Gene
Fama. Tom Gilovich was influenced by Amos Tversky. Carolyn Stein
and John Ezekowitz were *Moneyball*'s legacy. Matt and Max were in
their formative years when Curry was redefining their favorite sport,
and they were precisely the right age to be profoundly shaped by the
most transformational player of their lifetime. Curry was singularly
responsible for a fundamental shift in basketball strategy that fil-
tered down to every level of the sport. Matt and Max belonged to the
first wave of kids who followed his lead. It felt only right that Matt
and Max were taking the lessons of Stephen Curry, who benefited
from the hot hand as much as anyone in basketball, and bringing
them to Gilovich's experiment about the hot hand.

Matt and Max took their positions behind the three-point line.
This being a proper scientific experiment, Willie marked their chosen locations with tape. Matt and Max were then instructed to alternate taking one hundred shots.

"You're going to say before each shot whether you're feeling hot or
not," Willie said.

Matt and Max nodded in approval.

"You might be tempted in the beginning to exaggerate and think
you can hit 50 percent from here," Gilovich said. "Don't. Whatever

you realistically think your 50 percent shot is. Don't try to impress us by shooting from the parking lot."

Gilovich looked around the gym as Matt and Max got warm. The latest research about the hot hand had put him in a tricky position. Over the years he'd read all the papers that claimed that he was wrong about the hot hand, and he'd been able to brush them off with a shrug. But this new research was sticking, and the natural reaction for any tenured professor in his position would have been to stonewall other scholars. Instead he gave some of them his data and offered generous comments. He was trying to be a mensch about the whole thing because it takes a certain level of humility to reach the truth. There wasn't much he could do now about this paper that he'd written more than three decades earlier. He didn't have Ezekowitz and Stein's data back then, and it wasn't his fault that he'd missed Miller and Sanjurjo's bias. No one else had seen it, either. It was a bit like blaming everyone who came before Isaac Newton for not understanding the basic laws of physics.

The truth about the hot hand would eventually come out whether he liked it or not, and the least he could do was accelerate the process of epiphanies. He hadn't anticipated that he would be testing for the hot hand again when he published one of his first papers all those years ago. But here he was.

Everything had changed even though nothing had. The basketball gym was still a fantastic place for an experiment. Every game was a frenzy of enormous men pulling off insane feats of athleticism to put a piece of leather through a metal ring, but they didn't have to worry about any of that delirium during this session with Matt and Max. The whole point of studying the hot hand here was to apply the lessons of highly controlled environments to the sorts of places where anything goes.

There are certain environments where streaks exist. But there are also certain environments where streaks don't exist. Even more troublesome are the certain environments where streaks *might* exist but only if they're manipulated the right way.

Students like Matt, Max, and Willie would've been taught sometime in their tenure at Cornell that our tendency is to believe streaks exist. But one of the gratifying parts about growing up and figuring out our places in the world is recognizing where streaks exist, where they don't, and where they might. Do we dare to believe that we can break the bonds of logic? Or should we accept our limits and adhere to the basic laws of chance? The stakes of this natural pursuit of streaks are nothing short of existential. If you're Nick Hagen, you can lose the farm. But if you're Stephen Curry, you can win everything.

Once we begin to acknowledge the existence of the hot hand, or at least the possible existence of the hot hand, we can look around and find the three-point lines in our daily lives. Stephen Curry went from playing *NBA Jam* to playing in the NBA in the time it took for the hot hand to go from real to fake to maybe real again. Some of that had to do with incredible luck. If he'd come along a few years earlier or later, if he'd been drafted by another team, if he hadn't been late to the bus and if his bus hadn't gotten pulled over on the way to Madison Square Garden, then Curry might not have become one of the most influential basketball players ever. But he pounced when the opportunity presented itself.

Sometimes we can exploit changes in the rules of our controlled environments. Sometimes we have to manufacture the conditions for ourselves in uncontrolled environments. And sometimes the hot hand requires talent, circumstance, and the good fortune of having a number of things break our way. But the allure of streaks is so tantalizing that we naturally pour resources into facilitating the potential to get on a roll. It's part of who we are. We see them where they exist, where they don't, and where they might.

But there wasn't time to think about that now. Matt and Max were warm. They were ready for the experiment to begin.

Matt worked his way around the perimeter as Max stayed underneath the basket to rebound. He made a whole lot more shots than he missed. He didn't wait to make three shots in a row to say that he was hot. He felt that he was in the zone after one shot or two shots, and sometimes he even felt like he had the hot hand after missing a shot. He muttered yes and no—but mostly yes—loud enough for Willie to take notes on his clipboard and turn his yeses and noes into data.

"Three in a row," Gilovich marveled under his breath. "Four in a row. This guy's a good shooter!"

Matt kept shooting. He kept feeling hot.

"Yes," Matt said.

"Five," Gilovich said.

"Yes."

"Six."

"Yes."

"Seven. Wow!"

"Yes."

"Eight."

"Yes."

"Nine!" Gilovich said. "Ten of eleven. Eleven of twelve. Twelve of thirteen. Thirteen of fourteen. Fourteen of fifteen. Miller and Sanjurjo are going to be very happy to see this."

He wasn't mad. He was amused. He even sounded a little bit impressed.

"This is the best shooting of all the players we've had out here," Gilovich said.

By the time it was Max's turn to shoot, Matt had to leave for class, and that meant Max needed a rebounder. There were only three other people left in the gym. Willie had to write "yes" and "no" before every shot. That meant there were two people for the job: Tom Gilovich and me. We both volunteered to sprint around the court shagging rebounds and hoping our passes back to him were good enough that he would make his next shot and make us run less. Max made it through his three rounds of shots without breaking a sweat. I felt like I'd climbed Kilimanjaro. The sharp pain radiating throughout my entire lower body was the price of scientific inquiry.

We left the gym and walked back to the psychology building. There was a printed stack of paper on the coffee table in Gilovich's office. I looked down and noticed some familiar names: Josh Miller and Adam Sanjurjo. He'd been reading their study.

Gilovich made his way over to his desk, reached for another pile, and leafed through some more papers. It was a record of everything that Willie had been writing over the past few weeks. On his desk waiting to be coded and then decoded were hundreds of shots from basketball players like Matt and Max, along with their intuitions about whether they had the hot hand. This was his version of better data.

He was a long while away from knowing what it might say and which directions it might lead him. What to think about the hot hand was a question he was still trying to answer. But now it was getting late. I could tell it was time for me to go. I said goodbye knowing the only thing either of us could say with any certainty was that this discussion that went back decades would have to continue another day. The debate wasn't nearly over yet. Maybe it never will be. And isn't that the best part?

AN AUTHOR'S NOTE ON SOURCES

This book wouldn't exist without the work of so many people before me and the cooperation of so many people who spoke with me. I'm incredibly grateful to everyone in both groups.

Thank you to Kyle Allen and everyone associated with the Pine City Dragons. Sorry to everyone associated with my own basketball career.

I couldn't have picked a better time to start writing about the NBA, considering my first season covering the league happened to be the year of Stephen Curry's ascendance. I think he'll go down as one of the most influential athletes ever, and I've tried to reflect that in my stories about Curry and the Golden State Warriors for the *Journal*, which have been far too many to count. Chapter one is based on interviews with Curry and the people around him—Dell Curry, Warriors executives, and even his youth coaches. I'm in debt to the coverage of the local beat reporters around the team on a daily basis and the national basketball reporters who have chronicled Curry's unlikely rise. I remember exactly where I was on the night that Curry got hot. Unfortunately, it was not Madison Square Garden. I was able to read the coverage of Curry's theatrics and reconstruct that performance anyway. The sections in this chapter about Mark Turmell and the creation of *NBA Jam* are based on several long interviews with him. He was also gracious enough to provide me with access to his press clippings and some of the papers in his archives. Greg Voss's profile in *Softline* magazine was a useful look at Turmell when he was a teenager, and the interviews with Turmell by Jamin Warren in *Kill Screen Daily* and by Paul Drury in *Retro Gamer* gave me the background I needed to chat with him without sounding completely naive. *Sports Illustrated*'s oral history of *NBA Jam* by Alex Abnos and Dan Greene was a great read and a fantastic resource. I've never played *Bubble Safari*, but I feel like I have after reading Turmell's primer for his company's blog.

If there is anything you want to know about William Shakespeare, then James Shapiro is a pretty good person to ask. I couldn't have written the Shakespeare portions of chapter two without his book *The Year of Lear: Shakespeare in 1606*. It's an indispensable delight even (and maybe especially) if you don't know much about Shakespeare. The other book that paved the way for this chapter was *Politics, Plague, and Shakespeare's Theater* by J. Leeds Barroll, who writes with alarming clarity about the plague. Liane Curtis is responsible for almost everything the world now knows about Rebecca Clarke. Every quote from Clarke in this chapter comes from *A Rebecca Clarke Reader*, the collection of essays and interviews that Curtis edited. There have been millions of words already written about *The Princess Bride*, and I would read a few million more. The history of the movie was covered in an oral history in *Entertainment Weekly*, a retrospective in *Variety*, and a roundtable conducted by *The Hollywood Reporter*. Rob Reiner has answered hundreds if not thousands of questions about *The Princess Bride* over the course of his career. In addition to all the profiles of Reiner and interviews with Reiner, I found Cary Elwes's memoir, *As You Wish*, to be as delightful as his role in the movie.

I relied on many people to help craft my understanding of Stanford in the 1980s for chapter three, none more than Tom Gilovich, Bob Vallone, Lee Ross, and especially Barbara Tversky, whose comments I cherished. While there is no worse feeling as a writer to learn that Michael Lewis has written about people you're writing about, his typically brilliant book *The Undoing Project* is required reading about Daniel Kahneman and Amos Tversky. I used some of Lewis's reporting and research to buttress my own, which feels a bit like attaching a kayak to a yacht. Also essential was *The Essential Tversky*, a collection of his most influential papers. Kahneman's eulogy for Tversky was the moving tribute that he richly deserved. I'm also grateful to Kahneman for not running into the nearest cab when I introduced myself at Josh Miller's talk. In addition to my own interviews with Gilovich, I relied on two lengthy interviews he's given to Barry Ritholtz for his *Masters in Business* podcast and Alan Reifman for his blog *The Hot Hand in Sports*. Amos Tversky died long before I ever got a chance to badger him with questions about the hot hand, but many other reporters had the privilege of chatting with him, and I'm especially grateful to Kevin McKean for his pieces in *Discover*. Thanks to Stanford University's oral history project, the New York Public Library, and the Daniel Boone Regional Library. For the history of shuffle at Apple and Steve Jobs's reaction to the uproar, I relied on the reporting in Steven Levy's book *The Perfect Thing: How the iPod Shuffles Commerce, Culture, and Coolness* and his *Wired* essay, called "Requiem for the iPod Shuffle." Lukáš Poláček wrote about his algorithm better than I could and thoughtfully explained his work to me. To a few Spotify employees who asked not to be named but might be reading this: you helped enormously.

David Booth bought the rules of basketball, but only because of Josh Swade. His book *The Holy Grail of Hoops: One Fan's Quest to Buy the Original Rules of Basketball* is the definitive account of how Naismith's creation made its way back to Naismith Drive, and I took the dialogue from the auction from the ESPN 30 for 30 documentary *There's No Place Like Home*, directed by Swade and Maura Mandt. It's a reporter's dream to be a fly on the wall. He was. David

Booth didn't have to speak with me. I'm enormously grateful that he did, and I know it's only because of Alex Stockham. Booth isn't written about as often as other billionaires, but there have been several extensive profiles of him over the years, the best one by Jason Zweig of *The Wall Street Journal*. Thanks to the people who have not only interviewed Booth at length over the years but also uploaded those interviews to YouTube, where curious minds and desperate authors could watch them. Thanks to David Rubenstein for his candid recollections. Where to start with sugar beets! I'm endlessly appreciative of Nick Hagen and Molly Yeh for their hospitality, cooperation, and friendship. For everything you could ever want to know about sugar beets, I recommend a special episode of *America's Heartland* entirely about the harvest in Minnesota. Nick helpfully passed along a history of Hagen Farms from an article by Dan Looker in the March 2003 issue of *Successful Farming* and Bernt Hagen's entry in the *Compendium of History and Biography of Polk County,* and *Molly on the Range: Recipes and Stories from An Unlikely Life on a Farm* has the rare distinction of being an unlikely resource and one of my favorite cookbooks.

I read about Alaa Al-Saffar in a fascinating story in the *San Diego Union-Tribune* by Kate Morrissey, and he wouldn't be in chapter five of this book if not for Morrissey's initial article. Thank you to the Al-Saffar family for their grace. Syracuse University's Transactional Records Access Clearinghouse was a helpful resource of data. Judge Bruce Einhorn was generous with his time explaining asylum law. I couldn't have written about Jed Lowrie without Brooks Baseball's database or Fernando Alcala's assistance. I'm grateful to all the Texas Rangers beat writers who covered Justin Grimm's debut. And the *Off the Lip Radio Show* had incisive interviews with Bill Miller that helped me understand the life of an umpire.

The stories in chapter six are some of the most intimate and personally revealing in this book, and I can't properly express how much I appreciate the trust of everyone interviewed, including Ari Kaplan and especially Marvin Makinen. This book would be immeasurably worse if not for his knowledge and kindness. The inspiration to look into the Raoul Wallenberg saga came when I read Hillel Kuttler's fabulous story for *Tablet* with an irresistible headline: "Sabermetrician Ari Kaplan Uses the Science of Balls and Strikes to Illuminate the Fate of Holocaust Rescuer Raoul Wallenberg." Gal Oz, Miky Tamir, and especially Brian Kopp were instrumental in helping me understand the early years of SportVU. John Ezekowitz and Carolyn Stein were not only great pizza company but quite gracious in letting me write about their college papers. Of all the people who have written about their work, I found Zach Lowe's story in *Grantland* to be the one that clarified my thinking, which is yet another reason he's the world's best NBA writer.

Now, finally, chapter seven. Josh Miller and Adam Sanjurjo had endless patience and good humor explaining how coin flips worked, among many other things. I have known them for nearly five years now, and I have a feeling I'll be reading their work for the rest of my life. Andrew Gelman agreed to an interview in front of his classroom with students watching—which was a fantastic idea. Teio Meedendorp and Louis van Tilborgh provided me with their recollections of an unforgettable moment, and Don Johnson agreed to walk around the Museum of Fine Arts in Houston with me, which was a lovely way to spend a hot afternoon. I'm lucky to have written this book at a time when Van Gogh's letters are fully archived and annotated. Instead of digging through dusty old files in libraries and museums, they are now readily available, and that's because of the Van Gogh Museum.

I'm sure I have, but I hope I haven't forgotten anyone.

ENDNOTES

INTRODUCTION

6 **"In all honesty"**: Ben Cohen, "The Basketball Team That Never Takes a Bad Shot," *Wall Street Journal,* January 30, 2017.

7 **"That's how you"**: Ibid.

ONE: ON FIRE

12 **"The only problem"**: Greg Voss, "Sneaking Up on Success: An Interview with Mark Turmell," *Softline,* November 1981.

13 **"I kept plugging"**: Unless otherwise noted, quotes are from author interviews. See the Author's Note on Sources.

13 **Apple cofounder Steve Wozniak's company**: Steve Wozniak, *iWoz: Computer Geek to Cult Icon* (New York: W. W. Norton, 2007).

18 **his family's stucco**: David Fleming, "Stephen Curry: The Full Circle," *ESPN the Magazine,* April 23, 2015.

20 **"We thought the numbers"**: Alex Abnos and Dan Greene, "Boomshakalaka: The Oral History of *NBA Jam,*" *Sports Illustrated,* July 6, 2017.

26 **"If you want to exceed"**: John Hollinger, PER Diem, ESPN, March 27, 2009.

26 **"I love the three-point shot"**: Pete Carril, *The Smart Take from the Strong: The Basketball Philosophy of Pete Carril* (New York: Simon & Schuster, 1997), 133.

27 **"We refer to Bob"**: Emmanuelle Ejercito, "Everybody Loves Bob," *Daily Bruin,* February 27, 1997.

28 **"Were you serious"**: Tim Kawakami, "Bob Myers Interview: How the Warriors GM Was Hired [...]," *Talking Points* (blog), *Mercury News,* March 11, 2016.

28 **"What's really interesting"**: Ben Cohen, "The Golden State Warriors Have Revolutionized Basketball," *Wall Street Journal,* April 6, 2016.

29 **"I didn't even feel him"**: Jim Johnson, "Pacers Beat Warriors After 4th-Quarter Scuffle," Associated Press, February 26, 2013.

31 **"The way a long-distance swimmer"**: Mihaly Csikszentmihalyi, *Flow: The Psychology of Optimal Experience* (New York: Harper Perennial, 1991), 48.

32 **Queensway Christian College**: Lee Jenkins, "Stephen Curry's Next

Stage: MVP Has Warriors Closing in on the NBA Finals," *Sports Illustrated,* May 20, 2015.

33 **"There was nothing":** Frank Isola, "Stephen Curry Scores 54 Points at Garden [. . .]," *New York Daily News,* February 28, 2013.

35 **"I've never been":** Scott Fowler, "Curry Hits Broadway with Rare Performance," *Charlotte Observer,* March 1, 2013.

37 **Curry will tell you about three:** Kathleen Elkins, "NBA Star Stephen Curry Shares the 3 Moments When He Knew He'd 'Made It,'" CNBC, September 7, 2016.

TWO: THE LAW OF THE HOT HAND

42 **"a concentrated efflorescence":** J. Leeds Barroll, *Politics, Plague, and Shakespeare's Theater: The Stuart Years* (Ithaca, NY: Cornell University Press, 1991), 152.

43 **"as a composer":** Herbert F. Peyser, "Gifted Artists Join in Unique Recital," *Musical America,* 1918.

43 **"I thought it's idiotic":** Ellen D. Lerner, "Musicologist Ellen D. Lerner Interviews Rebecca Clarke, 1978 and 1979," in *A Rebecca Clarke Reader,* ed. Liane Curtis (Bloomington: Indiana University Press, 2004), 204.

43 **"It had much more attention":** Robert Sherman, "Robert Sherman Interviews Rebecca Clarke About Herself," in Curtis, *Rebecca Clarke Reader,* 172.

43 **"one should not overlook":** Hiram Kelly Moderwell, "Makers of Music," *Vogue,* April 15, 1918.

45 **"You should have *seen*":** Rebecca Clarke, "Rebecca Clarke's 1977 Program Note on the Viola Sonata," in Curtis, *Rebecca Clarke Reader,* 226.

45 **"I put my things away":** Lerner, "Musicologist," 205.

45 **"that one little whiff":** Sherman, "Interviews Rebecca Clarke," 171.

46 **"Why did you stop":** Ibid., 176.

46 **"There was a lot":** Curtis, introduction, *Rebecca Clarke Reader,* 4n5.

46 **"I didn't":** Sherman, "Interviews Rebecca Clarke," 176.

46 **"Every now and then":** Curtis, introduction, *Rebecca Clarke Reader,* 1.

47 **"There's nothing in the":** Sherman, "Interviews Rebecca Clarke," 176–77.

47 **"Most people don't":** Ibid., 77.

48 **"Had she not been":** Peter G. Davis, "Rewarding Program Assembled by Toby Appel for Viola Recital," *New York Times,* April 4, 1977.

49 **"I did my own":** Sherman, "Interviews Rebecca Clarke," 179.

50 **"one of the funniest":** Roger Ebert, review of *This Is Spinal Tap, Chicago Sun-Times,* March 1, 1985.

51 **"He is successful":** Ron Base, "Fathers of the Princess Bride," *Toronto Star,* September 26, 1987.

51 **"We love your films":** Susan King, "'The Princess Bride' Turns 30: Rob Reiner, Robin Wright, Billy Crystal Dish About Making the Cult Classic," *Variety,* September 25, 2017.

52 **"*The Princess Bride* is my favorite"**: Drew McWeeny, "The M/C Interview: Rob Reiner Talks 'Flipped,' 'Princess Bride,' 'Misery' and More," *HitFix*, August 4, 2010.

54 **The data they collected:** Lu Liu et al., "Hot Streaks in Artistic, Cultural, and Scientific Careers," *Nature* 559, no. 7714 (July 2018): 396–99.

57 **"Plague was the single":** Jonathan Bate, *Soul of the Age: A Biography of the Mind of William Shakespeare* (New York: Random House, 2009), 4.

62 **"This meant that his days":** James Shapiro, *The Year of Lear: Shakespeare in 1606* (New York: Simon & Schuster, 2015), 29.

63 **"We have no idea":** Ibid., 292.

THREE: SHUFFLE

65 **"The users were asking":** Lukáš Poláček, "How to Shuffle Songs?" *Spotify Labs*, February 28, 2014, labs.spotify.com/2014/02/28/how-to-shuffle-songs.

68 **"It really is random":** Steve Jobs Keynote, World Wide Developers Conference 2005, https://www.youtube.com/watch?v=B6iF6yTiNlw.

70 **"Our brain":** Dave Lee, "How Random Is Random on Your Music Player?" BBC News, February 19, 2015.

71 **"steeples of excellence":** C. Stewart Gillmor, *Fred Terman at Stanford: Building a Discipline, a University, and Silicon Valley* (Stanford, CA: Stanford University Press, 2004).

74 **"Those who have been":** See the section "Eulogy for Amos Tversky (June 5, 1996)," in Daniel Kahneman, "Biographical," Sveriges Riksbank Prize in Economic Sciences in Memory of Alfred Nobel 2002, Nobel Prize, nobelprize .org/prizes/economic-sciences/2002/kahneman/biographical.

74 **The joke among them:** *David and Goliath: Underdogs, Misfits and the Art of Battling Giants* by Malcolm Gladwell (Little, Brown, 2013).

76 **"When the initial idea is good":** Afterword by Daniel Kahneman in *The Essential Tversky*, page 366.

76 **"Sometimes they lead":** Amos Tversky and Daniel Kahneman, "Judgment Under Uncertainty: Heuristics and Biases," *Science* 185, no. 4157 (Sept. 27, 1974): 1124.

77 **"Consider a hypothetical scientist":** Amos Tversky and Daniel Kahneman, "Belief in the Law of Small Numbers," *Psychological Bulletin* 76, no. 2 (1971): 106.

78 **The most famous example:** *How We Know What Isn't So: The Fallibility of Human Reason in Everyday Life* by Thomas Gilovich (Free Press, 1991), 19.

78 **"People see patterns":** Kevin McKean, "The Orderly Pursuit of Pure Disorder," *Discover*, January 1987.

79 **"Amos had simply":** Kahneman, "Biographical."

83 **"glitch in the system":** Andreas Wilke and H. Clark Barrett, "The Hot Hand Phenomenon as Cognitive Adaptation to Clumped Resources," *Evolution and Human Behavior* 30, no. 3 (May 2009): 161–69.

86　**The three monkeys sipped:** Tommy C. Blanchard, Andreas Wilke, and Benjamin Y. Hayden "Hot-Hand Bias in Rhesus Monkeys," *Journal of Experimental Psychology: Animal Learning and Cognition* 40, no. 3 (July 2014): 280–86.

91　**"Very often the search":** James Gleick, "'Hot Hands' Phenomenon: A Myth?" *New York Times*, April 19, 1988.

91　**"The present data":** Thomas Gilovich, Robert Vallone, and Amos Tversky, "The Hot Hand in Basketball: On the Misperception of Random Sequences," *Cognitive Psychology* 17, no. 3 (July 1985): 313.

92　**"Who is this guy?":** "High-Handed Professor's Comments Called Hot Air," *USA Today*, August 30, 1985.

92　**"There are plenty":** Kevin McKean, "When You're Hot, You're Not," *Discover*, June 1985.

93　**Only a few months earlier:** Sylvia Nasar, *A Beautiful Mind: A Biography of John Forbes Nash Jr.* (New York: Simon & Schuster, 1998), 372–73.

94　**"I've been in":** McKean, "Orderly Pursuit."

FOUR: BET THE FARM

95　**"You spend *too* much time":** James Naismith, *Basketball: Its Origin and Development* (Lincoln: University of Nebraska Press, 1996), 21.

96　**"I thought there":** Ibid., 23.

96　**"I made up my mind":** Ibid.

96　**"Now would be a good time":** Ibid., 37.

97　**"How I hated the thought":** Ibid., 42.

97　**"I have two old peach baskets":** Ibid., 53.

98　**"Why not call it basketball?":** Ibid., 60.

98　**"For many years":** Eugene F. Fama, "The Behavior of Stock-Market Prices," *Journal of Business* 38, no. 1 (Jan. 1965): 34.

102　*What's wrong here?:* David Booth speech and Q&A, VIP Distinguished Speaker Series, McCombs School of Business, University of Texas, Austin, February 26, 2013, youtu.be/cCp1m7rG0Q0.

102　**"I realized they":** Ibid.

102　**"One of the first things":** Ibid.

103　**"a country bumpkin":** Molly Yeh, *Molly on the Range: Recipes and Stories from an Unlikely Life on the Farm* (New York: Rodale, 2016), 18.

108　**"lunatic monster":** Ibid, 133.

108　**When his first farm:** R. I. Holcombe and William H. Bingham, eds., "Bernt Hagen," *Compendium of History and Biography of Polk County, Minnesota* (Minneapolis, MN: W. H. Bingham, 1916), 312–13.

111　**removed the sauna:** Vanessa Sumo, "The Science," *Chicago GSB Magazine*, Winter 2009: 18.

111　**Rex Sinquefield was:** Shawn Tully, "How the Really Smart Money Invests," *Fortune*, July 6, 1998.

112 **"I'd compare stock-pickers"**: Ibid.

113 **"The first time"**: Booth speech and Q&A, VIP Distinguished Speaker Series.

114 **"One orangutan"**: Lydialyle Gibson, "Return on Principles," *University of Chicago Magazine*, January–February 2009.

115 **"Most MBAs think"**: Ibid.

115 **One paper looked at:** Andrew Mauboussin and Samuel Arbesman, "Differentiating Skill and Luck in Financial Markets with Streaks," *SSRN* (2011): dx.doi.org/10.2139/ssrn.1664031.

116 **"Most professions have value"**: Warren Buffett's comments at the 2006 Berkshire Hathaway annual meeting are found in CNBC's Warren Buffett Archive, buffett.cnbc.com/video/2006/05/06/morning-session---2006-berkshire-hathaway-annual-meeting.html.

116 **"What followed was the sound"**: Warren Buffett wrote about the bet in Berkshire Hathaway's 2016 annual report, 22 (available at berkshirehathaway.com/2016ar/2016ar.pdf).

117 **"I've often been asked"**: Ibid., 24–25.

117 **"Ignore the chatter"**: Warren Buffet, "To the Shareholders of Berkshire Hathaway Inc.:" https://www.berkshirehathaway.com/letters/2013ltr.pdf.

117 **"We would only be"**: David Booth interview by James K. Glassman, George W. Bush Presidential Center, September 18, 2012, bushcenter.imgix.net/legacy/Tax_Competition_and_4percent_Growth_09-18-12_Chicago_0.pdf.

118 **He works from:** Edward Lewine, "There's a Method to My Desk's Madness," *New York Times*, May 18, 2013.

118 **"largely unknown"**: Robert A. Guth, "Chicago Business School Gets Huge Gift," *Wall Street Journal*, November 7, 2008.

119 **"the most significant"**: The descriptions of the auction lots are taken from Sotheby's catalog.

119 **Redden was used:** James Barron, "He's Auctioned the 1776 Declaration, Twice," *New York Times*, July 4, 2000.

119 **"I don't think"**: Christopher Michaud, "Magna Carta Fetches $21.3 Million at Sotheby's Auction," Reuters, December 18, 2007.

119 **"I buy everything"**: Julie Segal, "David Rubenstein's Monopoly Money," *Institutional Investor*, May 4, 2017.

120 **"The third and final sale"**: *There's No Place Like Home*, dirs. Maura Mandt and Josh Swade, *30 for 30*, ESPN Films, 2012.

FIVE: WHEEL OF FORTUNE

127 **chiseling stone:** Gabriel Gatehouse, "Baghdad Diary: Saddam's Sculptor Makes Comeback," BBC News, June 16, 2010.

127 **"They called me an infidel"**: Ibid.

130 Mejdal studied aeronautical: Sig Mejdal was a character in Sam Walker's *Fantasyland: A Season on Baseball's Lunatic Fringe* (New York: Viking, 2006), and the story of his scouting trip to see Jed Lowrie was informed by Ben Reiter's *Astroball: The New Way to Win It All* (New York: Crown, 2018).

132 He made some extra cash: Bill Miller has been profiled a few times and repeated the same details of his life story. Among the sources I used were Scott Fields, "The Boy of Summer," *UCLA Magazine,* October 1, 2010, and Bruce Weber, *As They See 'Em: A Fan's Travels in the Land of Umpires* (New York: Scribner, 2009).

132 he especially loved: Miller has given long interviews to *Off the Lip Radio Show* in his native Santa Cruz, available at offthelipradio.com/live-stream.

133 He was a friend: Jeff Sullivan, "Incredulous Responses to Bill Miller's Strike Zone," FanGraphs, October 30, 2017, blogs.fangraphs.com/incredulous-responses-to-bill-millers-strike-zone.

133 "High pitches, low pitches": Michael Lopez and Sadie Lewis, "An Exploration of MLB Umpires' Strike Zones," *Hardball Times,* May 4, 2018, tht.fangraphs.com/an-exploration-of-mlb-umpires-strike-zones.

134 It should have been: PITCHf/x data for every pitch of every game is available on the website Brooks Baseball (brooksbaseball.net).

134 They manage to nail: Daniel Chen, Tobias J. Moskowitz, and Kelly Shue, "Decision Making Under the Gambler's Fallacy: Evidence from Asylum Judges, Loan Officers, and Baseball Umpires," *Quarterly Journal of Economics* 131, no. 3 (Aug. 2016): 1181–242.

137 When they set about: Rachel Croson and James Sundali, "The Gambler's Fallacy and the Hot Hand: Empirical Data from Casinos," *Journal of Risk and Uncertainty* 30, no. 3 (2005): 195–209.

138 "They place significantly": Ibid., 205.

139 In one chapter: Pierre-Simon Laplace, *Philosophical Essay on Probabilities,* trans. Andrew I. Dale (1814; New York: Springer, 1995).

139 Peter Ayton and Ilan Fischer: Peter Ayton and Ilan Fischer, "The Hot Hand Fallacy and the Gambler's Fallacy: Two Faces of Subjective Randomness?" *Memory & Cognition* 32, no. 8 (Dec. 2004): 1369–78.

140 "Any belief": Peter Ayton, "Fallacy Football," *New Scientist,* September 19, 1998.

143 "People seem to believe": Ayton and Fischer, "Hot Hand Fallacy," 1370.

143 It was the most comprehensive: Jaya Ramji-Nogales, Andrew I. Schoenholtz, and Philip G. Schrag, "Refugee Roulette: Disparities in Asylum Adjudication," *Stanford Law Review* 60, no. 2 (Nov. 2007): 295–411.

143 "Whether the asylum applicant": Ibid., 378.

144 "There is remarkable": Ibid., 302.

151 The real number: Sisi Wei and Nick Fortugno, "The Waiting Game," *ProPublica* and WNYC, April 23, 2018.

151 **There were 320,663 people:** These figures were taken from July 2018 data published by the U.S. Citizenship and Immigration Services.

SIX: THE FOG

155 **"To a traveler":** Albin E. Johnson, "What I Saw in Sweden," *Rotarian*, September 1944.

156 **"That," he said:** Ray Furlong, "Wallenberg family mark centenary with plea for truth," BBC News, August 8, 2012, https://www.bbc.com/news/world-europe-19101339.

156 **"I will be happy":** Ingrid Carlberg, *Raoul Wallenberg: The Biography*, trans. Ebba Segerberg (London: MacLehose Press, 2015), 44.

156 **"The conviction here":** Raoul Wallenberg, *Letters and Dispatches, 1924–1944*, trans. Kjersti Board (New York: Arcade, 2011), 111.

156 **"Hitchhiking gives you":** Ibid., 69.

157 **"sit around saying":** Carlberg, *Raoul Wallenberg*, 110.

157 **"We should find a person":** Ibid., 203.

158 **"Every day costs":** John Bierman, *Righteous Gentile: The Story of Raoul Wallenberg, Missing Hero of the Holocaust* (Toronto: Bantam Books, 1983), 36.

158 **"I think I've got":** Per Anger oral history interview by Joan Ringelheim, United States Holocaust Memorial Museum Collection, January 19, 1995, collections.ushmm.org/search/catalog/irn504796.

159 **And then so many:** Bierman, *Righteous Gentile*.

159 **"Not even the name":** Carlberg, *Raoul Wallenberg*, 280.

159 **"If anyone is capable":** Ibid., 365.

160 **He packed twenty thousand:** Per Anger oral history interview.

160 **He was so tireless:** Kati Marton, *Wallenberg: Missing Hero* (New York: Arcade, 1982).

160 **He hurled china:** Ibid.

160 **"have that Jew-dog Wallenberg shot":** Carlberg, *Raoul Wallenberg*, 386.

161 **"I admit that":** Bierman, *Righteous Gentile*, 100.

161 **"Of course it gets":** Per Anger oral history interview.

161 **"I do not know":** Carlberg, *Raoul Wallenberg*, 439.

170 **"Dear Mrs. Roosevelt":** Maj von Dardel letter to Eleanor Roosevelt, November 30, 1946, Franklin D. Roosevelt Presidential Library and Museum, fdrlibrary.marist.edu/_resources/images/ergen/ergen1367.pdf.

171 **"Not knowing is the worst":** Carlberg, *Raoul Wallenberg*, 634.

171 **Makinen was on the path:** Stu Borman, "A Chemistry Spy Story," *Chemical & Engineering News*, February 18, 2013.

173 **As he flipped:** Elenore Lester and Frederick E. Werbell, "The Lost Hero of the Holocaust: The Search for Sweden's Raoul Wallenberg," *New York Times Magazine*, March 30, 1980.

174 **"One can't accept":** Carlberg, *Raoul Wallenberg*, 630.

175 **"To hinder the investigation"**: Marvin W. Makinen and Ari D. Kaplan, Cell Occupancy Analysis of Korpus 2 of the Vladimir Prison, Swedish-Russian Working Group on the Fate of Raoul Wallenberg, December 15, 2000, raoul-wallenberg.eu/wp-content/uploads/2010/02/makinen_kaplan_report_pp01-16.pdf.

176 **He struck out:** Scott Harris, "Caltech Student Has the Stats to Make It to the Major Leagues," *Los Angeles Times,* June 6, 1990.

177 **Back then he was:** Michael Lewis, "The King of Human Error," *Vanity Fair,* December 2011.

177 **"Our ability to":** Ibid.

179 **"He complained":** Marvin W. Makinen and Ari D. Kaplan, *Cell Occupancy Analysis of Korpus 2 of the Vladimir Prison,* Swedish-Russian Working Group on the Fate of Raoul Wallenberg, December 15, 2000, raoul-wallenberg.eu/wp-content/uploads/2010/02/makinen_kaplan_report_pp01-16.pdf.

182 **"We called him that":** Josyp Terelya, *Josyp Terelya: Witness to Apparitions and Persecution in the USSR* (Milford, OH: Faith Publications, 1991), 132.

187 **"strange game":** Allan Shen, "Renowned Mathematician and Professor Elias Stein Passes Away at 87," *Daily Princetonian,* February 5, 2019.

188 **"How about":** Ben Cohen, "Moneyball 2.0: Students in Harvard Club Prep to Be GMs," *ThePostGame,* February 24, 2011, thepostgame.com/features/201102/moneyball-20-students-harvard-club-prep-be-sports-gms.

189 **"The sort of people":** Michael Lewis, *Moneyball: The Art of Winning an Unfair Game* (New York: W. W. Norton, 2003), 130.

190 **"The more basketball":** Amos Tversky and Thomas Gilovich, "The Cold Facts About the 'Hot Hand' in Basketball," *Chance* 2, no. 1 (1989): 21.

191 **It was from Mark Cuban:** "The Most Unlikely College Basketball Result of 2010," *Harvard Sports Analysis Collective,* August 6, 2010, harvardsportsanalysis.org/2010/08/the-most-unlikely-college-basketball-result-of-2010.

192 **"Hope exams are":** John Ezekowitz email to Carolyn Stein, December 20, 2012.

197 **"The key is reading data":** Adam Davidson, "Boom, Bust or What?" *New York Times Magazine,* May 2, 2013.

197 **"Each player has":** Gilovich et al.: "The Hot Hand in Basketball."

199 **"All I could see":** Kirk Goldsberry, "DataBall," *Grantland,* February 6, 2014, grantland.com/features/expected-value-possession-nba-analytics.

201 **"If this was":** Bill James, "Underestimating the Fog," *Baseball Research Journal* 33 (2004): 29.

202 **"No one has made":** Ibid., 33.

202 **"Let's look again":** Ibid.

203 **It amounted to:** Andrew Bocskocsky, John Ezekowitz, and Carolyn Stein, "Heat Check: New Evidence on the Hot Hand in Basketball," *SSRN* (2014): dx.doi.org/10.2139/ssrn.2481494.

204 **"At the very least":** Ibid.

204 **"It's an impressive piece":** Larry Summers email to Carolyn Stein, February 27, 2014.

SEVEN: THE VAN GOGH IN THE ATTIC

207 **Mustad was the scion:** The bulk of the biographical information about Mustad comes from Louis van Tilborgh, Teio Meedendorp, and Oda van Maanen, "'Sunset at Montmajour': A Newly Discovered Painting by Vincent van Gogh," *Burlington Magazine*, 155, no. 1327 (Oct. 2013): 696–705.

208 **Pellerin's impressive art collection:** Alex Danchev, *Cézanne: A Life* (New York: Pantheon Books, 2012).

212 **managed to track down:** Richard J. Jagacinski, Karl M. Newell, and Paul D. Isaac, "Predicting the Success of a Basketball Shot at Various Stages of Execution," *Journal of Sport & Exercise Psychology* 1, no. 4 (1979): 301–10.

213 **They worked on:** Joshua B. Miller and Adam Sanjurjo, "Is It a Fallacy to Believe in the Hot Hand in the NBA Three-Point Contest?" IGIER Working Paper No. 548, *SSRN* (2015): dx.doi.org/10.2139/ssrn.2611987.

218 **"The zenith":** Ronald Pickvance, *Van Gogh in Arles* (New York: Metropolitan Museum of Art, 1984), 11.

219 **"I've seen lots":** Vincent van Gogh letter to Theo van Gogh, March 9, 1888.

219 **"I was on a stony heath":** Vincent van Gogh letter to Theo van Gogh, July 5, 1888.

220 **"Among them are many with":** Vincent van Gogh letter to Theo van Gogh, August 13, 1888.

223 **the proportion of heads:** Joshua B. Miller and Adam Sanjurjo, "Surprised by the Hot Hand Fallacy? A Truth in the Law of Small Numbers," *Econometrica* 86, no. 6 (Nov. 2018): 2019–47.

224 **"It may be that":** Gleick, "'Hot Hands' Phenomenon."

226 **"They've found something":** Jordan Ellenberg, "'Hot Hands' in Basketball Are Real" *Slate*, October 26, 2015.

229 **Don Johnson wasn't:** Mary Dahdouh, "He Wed Science with Art, Solving Mystery of 'Sunset at Montmajour,'" *Houston Chronicle*, September 23, 2013.

229 **Johnson's work was:** Louis van Tilborgh et al., "Weave Matching and Dating of Van Gogh's Painting: An Interdisciplinary Approach," *Burlington Magazine*, February 2012.

232 **The first thing:** Teio Meedendorp explained their process of authentication in a TEDx Talk called "Discovering Vincent van Gogh's *Sunset at Montmajour*" at the University of St. Andrews, available at youtu.be /SyzdA_dQjD0.

232 **They were able to confirm:** "How Do You Spot a Real Van Gogh?" *Economist*, September 24, 2013.

233 **"It is still incomprehensible":** Van Tilborgh, Meedendorp, and Van Maanen, "'Sunset at Montmajour.'"

234 **"New discovery Van Gogh":** Frederique Haanen email to Don Johnson, September 9, 2013.

236 **In a blog post:** Andrew Gelman, "Hey—Guess What? There Really Is a Hot Hand!" *Statistical Modeling, Causal Inference, and Social Science,* July 9, 2015, statmodeling.stat.columbia.edu/2015/07/09/hey-guess-what-there-really -is-a-hot-hand.

239 **"raised a valid":** Yosef Rinott and Maya Bar-Hillel, "Comments on a 'Hot Hand' Paper by Miller and Sanjurjo (2015)." *SSRN* (2015): dx.doi .org/10.2139/ssrn.2642450.

BIBLIOGRAPHY

BOOKS

Barroll, J. Leeds. *Politics, Plague, and Shakespeare's Theater: The Stuart Years.* Ithaca, NY: Cornell University Press, 1996.

Bate, Jonathan. *Soul of the Age: A Biography of the Mind of William Shakespeare.* New York: Random House, 2009.

Bierman, John. *Righteous Gentile: The Story of Raoul Wallenberg, Missing Hero of the Holocaust.* Toronto: Bantam Books, 1983.

Carlberg, Ingrid. *Raoul Wallenberg: The Biography.* Translated by Ebba Segerberg. London: MacLehose Press, 2015.

Carril, Pete. *The Smart Take from the Strong: The Basketball Philosophy of Pete Carril.* New York: Simon & Schuster, 1997.

Csikszentmihalyi, Mihaly. *Flow: The Psychology of Optimal Experience.* New York: Harper Perennial, 1991.

Curtis, Liane, ed. *A Rebecca Clarke Reader.* Bloomington: Indiana University Press, 2004.

Danchev, Alex. *Cézanne: A Life.* New York: Pantheon Books, 2012.

Ellenberg, Jordan. *How Not to Be Wrong: The Power of Mathematical Thinking.* New York: Penguin Press, 2014.

Elwes, Cary. *As You Wish: Inconceivable Tales from the Making of* The Princess Bride. New York: Touchstone, 2014.

Gelman, Andrew and Deborah Nolan. *Teaching Statistics: A Bag of Tricks.* Oxford: Oxford University Press, 2002.

Gillmor, C. Stewart. *Fred Terman at Stanford: Building a Discipline, a University, and Silicon Valley.* Stanford, CA: Stanford University Press, 2004.

Gilovich, Thomas. *How We Know What Isn't So: The Fallibility of Human Reason in Everyday Life.* New York: Free Press, 1991.

Gladwell, Malcolm. *Blink: The Power of Thinking Without Thinking.* New York: Little, Brown, 2005.

———. *David and Goliath: Underdogs, Misfits, and the Art of Battling Giants.* New York: Little, Brown, 2013.

———. *The Tipping Point: How Little Things Can Make a Big Difference.* Little, Brown, 2000.

Goldman, William. *The Princess Bride: S. Morgenstern's Classic Tale of True Love and High Adventure.* New York: Harcourt Brace Jovanovich, 1973.

———. *Which Lie Did I Tell? More Adventures in the Screen Trade.* New York: Pantheon Books, 2000.

Goldsberry, Kirk. *SprawlBall: A Visual Tour of the New Era of the NBA.* New York: Houghton Mifflin Harcourt, 2019.

Helvey, Jennifer. *Irises: Vincent van Gogh in the Garden.* Los Angeles: J. Paul Getty Museum, 2009.

Kahneman, Daniel. *Thinking, Fast and Slow.* New York: Farrar, Straus & Giroux, 2011.

Kahneman, Daniel, Paul Slovic, and Amos Tversky. *Judgment Under Uncertainty: Heuristic and Biases.* Cambridge: Cambridge University Press, 1982.

Laplace, Pierre-Simon. *Philosophical Essay on Probabilities.* Translated by Andrew I. Dale. New York: Springer, 1995.

Levy, Steven. *The Perfect Thing: How the iPod Shuffles Commerce, Culture, and Coolness.* New York: Simon & Schuster, 2006.

Lewis, Michael. *Moneyball: The Art of Winning an Unfair Game.* New York: W. W. Norton, 2003.

———. *The Undoing Project: A Friendship That Changed the World.* London: Allen Lane, 2017.

Marton, Kati. *Wallenberg: Missing Hero.* New York: Arcade, 1982.

Naifeh, Steven and Gregory White Smith. *Van Gogh: The Life.* New York: Random House, 2011.

Naismith, James. *Basketball: Its Origin and Development.* Lincoln: University of Nebraska Press, 1996.

Nasar, Sylvia. A Beautiful Mind: *A Biography of John Forbes Nash Jr.* New York: Simon & Schuster, 1998.

Pickvance, Ronald. *Van Gogh in Arles.* New York: Metropolitan Museum of Art, 1984.

Reich, Nancy B.,"Rebecca Clarke: An Uncommon Woman." *A Rebecca Clarke Reader,* edited by Liane Curtis. Somerville, MA: The Rebecca Clarke Society: https://www.rebeccaclarke.org/wp-content/uploads/2019/08/Pt-I-Ch-1-Nancy-B-Reich-on-Rebecca-Clarke.pdf.

Reifman, Alan. *Hot Hand: The Statistics Behind Sports' Greatest Streaks.* Washington, D.C.: Potomac Books, 2011.

Reiter, Ben. *Astroball: The New Way to Win It All.* New York: Crown, 2018.

Shapiro, James. *The Year of Lear: Shakespeare in 1606.* New York: Simon & Schuster, 2015.

Simmons, Bill. *The Book of Basketball: The NBA According to the Sports Guy.* New York: Ballantine/ESPN Books, 2010.

Swade, Josh. *The Holy Grail of Hoops: One Fan's Quest to Buy the Original Rules of Basketball.* New York: Skyhorse Publishing, 2013.

Terelya, Josyp. *Josyp Terelya: Witness to Apparitions and Persecution in the USSR.* Milford, OH: Faith Publications, 1991.

Thompson, Marcus, II. *Golden: The Miraculous Rise of Steph Curry*. New York: Touchstone, 2017.

Tromp, Henk. *A Real Van Gogh: How the Art World Struggles with Truth*. Amsterdam: Amsterdam University Press, 2010.

Tversky, Amos. *The Essential Tversky*. Edited by Eldar Shafir. Cambridge, MA: MIT Press, 2018.

Walker, Sam. *Fantasyland: A Season on Baseball's Lunatic Fringe*. New York: Viking, 2006.

Wallenberg, Raoul. *Letters and Dispatches, 1924–1944*. Translated by Kjersti Board. New York: Arcade, 2011.

Weber, Bruce. *As They See 'Em: A Fan's Travels in the Land of Umpires*. New York: Scribner, 2009.

Wozniak, Steve. *iWoz: Computer Geek to Cult Icon*. New York: W. W. Norton, 2007.

Yeh, Molly. *Molly on the Range: Recipes and Stories from an Unlikely Life on a Farm*. New York: Rodale, 2016.

ACADEMIC PUBLICATIONS

Albert, Jim. "Streaky Hitting in Baseball." *Journal of Quantitative Analysis in Sports* 4, no. 1 (2008): 1–32.

Albert, Jim, and Patricia Williamson. "Using Model/Data Simulations to Detect Streakiness." *American Statistician* 55, no. 1 (Feb. 2001): 41–50.

Arkes, Jeremy. "Do Gamblers Correctly Price Momentum in NBA Betting Markets?" *Journal of Prediction Markets* 5, no. 1 (2011): 31–50.

———. "The Hot Hand vs. Cold Hand on the PGA Tour." *International Journal of Sport Finance* 11, no. 2 (May 2016): 99–113.

———. "Misses in 'Hot Hand' Research." *Journal of Sports Economics* 14, no. 4 (Aug. 2013): 401–10.

———. "Revisiting the Hot Hand Theory with Free Throw Data in a Multivariate Framework." *Journal of Quantitative Analysis in Sports* 6, no. 1 (Jan. 2010): 1–12.

Attali, Yigal. "Perceived Hotness Affects Behavior of Basketball Players and Coaches." *Psychological Science* 24, no. 7 (July 2013): 1151–56.

Ayton, Peter, and Ilan Fischer. "The Hot Hand Fallacy and the Gambler's Fallacy: Two Faces of Subjective Randomness?" *Memory & Cognition* 32, no. 8 (Dec. 2004): 1369–78.

"Barbara Tversky: An Oral History." Conducted by Natalie Marine-Street. Stanford Historical Society Oral History Program, Stanford University Department of Special Collections and University Archives, 2017, purl.stanford.edu/dj690gf3172.

Bar-Eli, Michael, Simcha Avugos, and Markus Raab. "Twenty Years of 'Hot Hand' Research: Review and Critique." *Psychology of Sport and Exercise* 7, no. 6 (2006): 525–53.

Berry, Scott M. "Does 'the Zone' Exist for Home-Run Hitters?" *Chance* 12, no. 1 (1999): 51–56.

Blanchard, Tommy C., Andreas Wilke, and Benjamin Y. Hayden. "Hot-Hand Bias in Rhesus Monkeys." *Journal of Experimental Psychology: Animal Learning and Cognition* 40, no. 3 (July 2014): 280–86.

Bocskocsky, Andrew, John Ezekowitz, and Carolyn Stein. "Heat Check: New Evidence on the Hot Hand in Basketball." *SSRN* (2014): dx.doi.org/10.2139/ssrn.2481494.

Bondt, Werner P. M. De. "Betting on Trends: Intuitive Forecasts of Financial Risk and Return." *International Journal of Forecasting* 9, no. 3 (Nov. 1993): 355–71.

Booth, David. Interview by James K. Glassman. George W. Bush Presidential Center, Dallas, TX, September 18, 2012, bushcenter.imgix.net/legacy/Tax_Competition _and_4percent_Growth_09-18-12_Chicago_0.pdf.

Boynton, David M. "Superstitious Responding and Frequency Matching in the Positive Bias and Gambler's Fallacy Effects." *Organizational Behavior and Human Decision Processes* 91, no. 2 (2003): 119–27.

Brown, William O., and Raymond D. Sauer. "Does the Basketball Market Believe in the Hot Hand? Comment." *American Economic Review* 83, no. 5 (Dec. 1993): 1377–86.

Burns, Bruce D. "Heuristics as Beliefs and as Behaviors: The Adaptiveness of the 'Hot Hand.'" *Cognitive Psychology* 48, no. 3 (May 2004): 295–331.

Burns, Bruce D., and Bryan Corpus. "Randomness and Inductions from Streaks: 'Gambler's Fallacy' Versus 'Hot Hand.'" *Psychonomic Bulletin & Review* 11, no. 1 (Feb. 2004): 179–84.

Camerer, Colin F. "Does the Basketball Market Believe in the 'Hot Hand'?" *American Economic Review* 79, no. 5 (Dec. 1989): 1257–61.

Caruso, Eugene M., Adam Waytz, and Nicholas Epley. "The Intentional Mind and the Hot Hand: Perceiving Intentions Makes Streaks Seem Likely to Continue." *Cognition* 116, no. 1 (July 2010): 149–53.

Castel, Alan D., Aimee Drolet Rossi, and Shannon McGillivray. "Beliefs About the 'Hot Hand' in Basketball Across the Adult Life Span." *Psychology and Aging* 27, no. 3 (Sept. 2012): 601–5.

Chen, Daniel, Tobias J. Moskowitz, and Kelly Shue. "Decision Making Under the Gambler's Fallacy: Evidence from Asylum Judges, Loan Officers, and Baseball Umpires." *Quarterly Journal of Economics* 131, no. 3 (Aug. 2016): 1181–242.

Clark, Russell D. "An Analysis of Streaky Performance on the LPGA Tour." *Perceptual and Motor Skills* 97, no. 5 (Oct. 2003): 365–70.

———. "Examination of Hole-to-Hole Streakiness on the PGA Tour." *Perceptual and Motor Skills* 100, no. 3 (June 2005): 806–14.

———. "An Examination of the 'Hot Hand' in Professional Golfers." *Perceptual and Motor Skills* 101, no. 3 (Dec. 2005): 935–42.

Cotton, Christopher, and Joseph Price. "The Hot Hand, Competitive Experience, and Performance Differences by Gender." *SSRN* (2006): dx.doi.org/10.2139 /ssrn.933677.

Croson, Rachel, and James Sundali. "The Gambler's Fallacy and the Hot Hand: Empirical Data from Casinos." *Journal of Risk and Uncertainty* 30, no. 3 (2005): 195–209.

Csapo, Peter, et al. "The Effect of Perceived Streakiness on the Shot-Taking Behaviour of Basketball Players." *European Journal of Sport Science* 15, no. 7 (2014): 647–54.

———. "How Should 'Hot' Players in Basketball Be Defended? The Use of Fast-and-

Frugal Heuristics by Basketball Coaches and Players in Response to Streakiness." *Journal of Sports Sciences* 33, no. 15 (2015): 1580–88.

Csapo, Peter, and Markus Raab. "'Hand Down, Man Down.' Analysis of Defensive Adjustments in Response to the Hot Hand in Basketball Using Novel Defense Metrics." *PLoS ONE* 9, no. 12 (2014): doi.org/10.1371/journal.pone.0114184.

DeLong, J. Bradford, et al. "Positive Feedback Investment Strategies and Destabilizing Rational Speculation." *Journal of Finance* 45, no. 2 (June 1990): 379–95.

Dohmen, Thomas, et al. "Biased Probability Judgment: Evidence of Incidence and Relationship to Economic Outcomes from a Representative Sample." *Journal of Economic Behavior & Organization* 72, no. 3 (Dec. 2009): 903–15.

Dorsey-Palmateer, Reid, and Gary Smith. "Bowlers' Hot Hands." *American Statistician* 58, no. 1 (Feb. 2004): 38–45.

Durbach, Ian N., and Jani Thiart. "On a Common Perception of a Random Sequence in Cricket: Application." *South African Statistical Journal* 41, no. 2 (Jan. 2007): 161–87.

Durham, Gregory R., Michael G. Hertzel, and J. Spencer Martin. "The Market Impact of Trends and Sequences in Performance: New Evidence." *Journal of Finance* 60, no. 5 (Oct. 2005): 2551–69.

Fama, Eugene F. "The Behavior of Stock-Market Prices." *Journal of Business* 38, no. 1 (Jan. 1965): 34.

Falk, Ruma. "The perception of randomness." Proceedings of the Fifth International Conference for the Psychology of Mathematics Education. Grenoble, France, 1981.

Filho, Edson Soares Medeiros, Luiz Carlos Moraes, and Gershon Tenenbaum. "Affective and Physiological States During Archery Competitions: Adopting and Enhancing the Probabilistic Methodology of Individual Affect-Related Performance Zones (IAPZs)." *Journal of Applied Sport Psychology* 20, no. 4 (2008): 441–56.

Fischer, Ilan, and Lior Savranevski. "Extending the Two Faces of Subjective Randomness: From the Gambler's and Hot-Hand Fallacies Toward a Hierarchy of Binary Sequence Perception." *Memory & Cognition* 43, no. 7 (Oct. 2015): 1056–70.

Gao, Shan, et al. "Second Language Feedback Abolishes the 'Hot Hand' Effect During Even-Probability Gambling." *Journal of Neuroscience* 35, no. 15 (Apr. 2015): 5983–89.

Gilden, David L., and Stephanie Gray Wilson. "On the Nature of Streaks in Signal Detection." *Cognitive Psychology* 28, no. 1 (Feb. 1995): 17–64.

———. "Streaks in Skilled Performance." *Psychonomic Bulletin & Review* 2, no. 2 (June 1995): 260–65.

Gilovich, Thomas, Robert Vallone, and Amos Tversky. "The Hot Hand in Basketball: On the Misperception of Random Sequences." *Cognitive Psychology* 17, no. 3 (July 1985): 295–314.

Gould, Stephen Jay. "The Streak of Streaks." *Chance* 2, no. 2 (1989): 10–16.

Gray, Rob, and Jonathan Allsop. "Interactions Between Performance Pressure, Performance Streaks, and Attentional Focus." *Journal of Sport & Exercise Psychology* 35, no. 4 (Aug. 2013): 368–86.

Green, Brett, and Jeffrey Zwiebel. "The Hot-Hand Fallacy: Cognitive Mistakes or Equilibrium Adjustments? Evidence from Major League Baseball." *Management Science* 64, no. 11 (Nov. 2018): 4967–5460.

Gronchi, Giorgio, and Steven A. Sloman. "Do Causal Beliefs Influence the Hot-Hand and the Gambler's Fallacy?" Proceedings of the Thirtieth Annual Conference of the Cognitive Science Society, Washington, D.C., 2008, 1164–68.

Gula, Bartosz, and Markus Raab. "Hot Hand Belief and Hot Hand Behavior: A Comment on Koehler and Conley." *Journal of Sport & Exercise Psychology* 26, no. 1 (2004): 167–70.

Guryan, Jonathan, and Melissa S. Kearney. "Gambling at Lucky Stores: Empirical Evidence from State Lottery Sales." *American Economic Review* 98, no. 1 (Mar. 2008): 458–73.

Hales, Steven D. "An Epistemologist Looks at the Hot Hand in Sports." *Journal of the Philosophy of Sport* 26, no. 1 (1999): 79–87.

Iso-Ahola, Seppo E., and Charles O. Dotson. "Psychological Momentum: Why Success Breeds Success." *Review of General Psychology* 18, no. 1 (Mar. 2014): 19–33.

Jagacinski, Richard J., Karl M. Newell, and Paul D. Isaac. "Predicting the Success of a Basketball Shot at Various Stages of Execution." *Journal of Sport & Exercise Psychology* 1, no. 4 (1979): 301–10.

Jagannathan, Ravi, Alexey Malakhov, and Dmitry Novikov. "Do Hot Hands Exist Among Hedge Fund Managers? An Empirical Evaluation." *Journal of Finance* 65, no. 1 (Feb. 2010): 217–55.

James, Bill. "Underestimating the Fog." *Baseball Research Journal* 33 (2004): 29–33.

Jetter, Michael, and Jay K. Walker. "Game, Set, and Match: Do Women and Men Perform Differently in Competitive Situations?" *Journal of Economic Behavior & Organization* 119 (Nov. 2015): 96–108.

Ji, Li-Jun, et al. "Culture and Gambling Fallacies." *SpringerPlus* 4, no. 1 (Sept. 2015): 510.

Kahneman, Daniel, and Mark W. Riepe. "Aspects of Investor Psychology." *Journal of Portfolio Management* 24, no. 4 (Summer 1998): 52–65.

Kennedy, Patrick, David B. Miele, and Janet Metcalfe. "The Cognitive Antecedents and Motivational Consequences of the Feeling of Being in the Zone." *Consciousness and Cognition* 30 (Nov. 2014): 48–61.

Klaassen, Franc J. G. M., and Jan R. Magnus. "Are Points in Tennis Independent and Identically Distributed? Evidence from a Dynamic Binary Panel Data Model." *Journal of the American Statistical Association* 96, no. 454 (June 2001): 500–509.

Koehler, Jonathan J., and Caryn A. Conley. "The 'Hot Hand' Myth in Professional Basketball." *Journal of Sport & Exercise Psychology* 25, no. 2 (2003): 253–59.

Köppen, Jörn, and Markus Raab. "The Hot and Cold Hand in Volleyball: Individual Expertise Differences in a Video-Based Playmaker Decision Test." *Sport Psychologist* 26, no. 2 (2012): 167–85.

Korb, Kevin B., and Michael Stillwell. "The Story of the Hot Hand: Powerful Myth or Powerless Critique." International Conference on Cognitive Science, Sydney, 2003.

Larkey, Patrick D., Richard A. Smith, and Joseph B. Kadane. "It's Okay to Believe in the 'Hot Hand.'" *Chance* 2, no. 4 (1989): 22–30.

Lauer, Michael S. "From Hot Hands to Declining Effects: The Risks of Small Numbers." *Journal of the American College of Cardiology* 60, no. 1 (July 2012): 72–74.

Levy, Haim, and Moshe Levy. "Overweighing Recent Observations: Experimental Results and Economic Implications." In *Marketing, Accounting and Cognitive Perspectives*, edited by Rami Zwick and Amnon Rapoport, 155–183. Vol. 3 of *Experimental Business Research*. Boston: Springer, 2005.

Liu, Lu, et al. "Hot Streaks in Artistic, Cultural, and Scientific Careers." *Nature* 559, no. 7714 (July 2018): 396–99.

Livingston, Jeffrey A. "The Hot Hand and the Cold Hand in Professional Golf." *Journal of Economic Behavior & Organization* 81, no. 1 (Jan. 2012): 172–84.

Ma, Xiao, Seung Hyun Kim, and Sung S. Kim. "Online Gambling Behavior: The Impacts of Cumulative Outcomes, Recent Outcomes, and Prior Use." *Information Systems Research* 25, no. 3 (Sept. 2014): 511–27.

Mace, F. Charles, et al. "Behavioral Momentum in College Basketball." *Journal of Applied Behavior Analysis* 25, no. 3 (Fall 1992): 657–63.

MacMahon, Clare, Jörn Köppen, and Markus Raab. "The Hot Hand Belief and Framing Effects." *Research Quarterly for Exercise and Sport* 85, no. 3 (Sept. 2014): 341–350.

Mauboussin, Andrew, and Samuel Arbesman. "Differentiating Skill and Luck in Financial Markets with Streaks." *SSRN* (2011): dx.doi.org/10.2139/ssrn.1664031.

Miller, Joshua B., and Adam Sanjurjo. "A Bridge from Monty Hall to the Hot Hand: The Principle of Restricted Choice." *Journal of Economic Perspectives* 33, no. 3 (Summer 2019): 144–62.

———. "A Cold Shower for the Hot Hand Fallacy: Robust Evidence that Belief in the Hot Hand Is Justified." IGIER Working Paper No. 518, *SSRN* (2014): dx.doi.org/10.2139/ssrn.2450479.

———. "Is It a Fallacy to Believe in the Hot Hand in the NBA Three-Point Contest?" IGIER Working Paper No. 548, *SSRN* (2015): dx.doi.org/10.2139/ssrn.2611987.

———. "A Primer and Frequently Asked Questions for 'Surprised by the Gamblers and Hot Hand Fallacies? A Truth in the Law of Small Numbers' (Miller and Sanjurjo 2015)." *SSRN* (2016): dx.doi.org/10.2139/ssrn.2728151.

———. "Surprised by the Hot Hand Fallacy? A Truth in the Law of Small Numbers." *Econometrica* 86, no. 6 (Nov. 2018): 2019–47.

———. "A Visible (Hot) Hand? Expert Players Bet on the Hot Hand and Win." *SSRN* (2017): dx.doi.org/10.2139/ssrn.3032826.

Miller, Joshua B., and Andrew Gelman. "Laplace's Theories of Cognitive Illusions, Heuristics, and Biases." *SSRN* (2018): dx.doi.org/10.2139/ssrn.3149224.

Miller, Steve, and Robert Weinberg. "Perceptions of Psychological Momentum and Their Relationship to Performance." *Sport Psychologist* 5, no. 3 (1991): 211–22.

Miyoshi, Hiroto. "Is the 'Hot-Hands' Phenomenon a Misperception of Random Events?" *Japanese Psychological Research* 42, no. 2 (May 2000): 128–33.

Morrison, Donald G., and David C. Schmittlein. "It Takes a Hot Goalie to Raise the Stanley Cup." *Chance* 11, no. 1 (1998): 3–7.

Narayanan, Sridhar, and Puneet Manchanda. "An Empirical Analysis of Individual Level Casino Gambling Behavior." *Quantitative Marketing & Economics* 10, no. 1 (Mar. 2011): 27–62.

Oskarsson, An T., et al. "What's Next? Judging Sequences of Binary Events." *Psychological Bulletin* 135, no. 2 (Mar. 2009): 262–85.

Parsons, Stephanie, and Nicholas Rohde. "The Hot Hand Fallacy Re-Examined: New Evidence from the English Premier League." *Applied Economics* 47, no. 4 (2014): 346–57.

Powdthavee, Nattavudh, and Yohanes E. Riyanto. "Would You Pay for Transparently Useless Advice? A Test of Boundaries of Beliefs in the Folly of Predictions." *Review of Economics and Statistics* 97, no. 2 (May 2015): 257–72.

Raab, Markus, Bartosz Gula, and Gerd Gigerenzer. "The Hot Hand Exists in Volleyball and Is Used for Allocation Decisions." *Journal of Experimental Psychology: Applied* 18, no. 1 (Mar. 2012): 81–94.

Rabin, Matthew. "Inference by Believers in the Law of Small Numbers." *Quarterly Journal of Economics* 117, no. 3 (Aug. 2002): 775–816.

Rabin, Matthew, and Dimitri Vayanos. "The Gambler's and Hot-Hand Fallacies: Theory and Applications." *Review of Economic Studies* 77, no. 2 (Apr. 2010): 730–78.

Ramji-Nogales, Jaya, Andrew I. Schoenholtz, and Philip G. Schrag. "Refugee Roulette: Disparities in Asylum Adjudication." *Stanford Law Review* 60, no. 2 (Nov. 2007): 295–411.

Rao, Justin M. "Experts' Perceptions of Autocorrelation: The Hot Hand Fallacy Among Professional Basketball Players." Working paper, 2009. Available at pdfs.semanticscholar.org/2f15/0b534bfa31a2c439eb9b0ebc2051faae8d9b.pdf.

Redelmeier, Donald A. and Amos Tversky. "Discrepancy Between Medical Decisions for Individual Patients and for Groups." *New England Journal of Medicine* 322, no. 16 (1990): 1162–64.

Rinott, Yosef, and Maya Bar-Hillel. "Comments on a 'Hot Hand' Paper by Miller and Sanjurjo (2015)." *SSRN* (2015): dx.doi.org/10.2139/ssrn.2642450.

Roney, Christopher J. R., and Lana M. Trick. "Sympathetic Magic and Perceptions of Randomness: The Hot Hand Versus the Gambler's Fallacy." *Thinking & Reasoning* 15, no. 2 (2009): 197–210.

Schilling, Mark F. "Does Momentum Exist in Competitive Volleyball?" *Chance* 22, no. 4 (2009): 29–35.

Shea, Stephen. "In Support of a Hot Hand in Professional Basketball and Baseball." *PsyCh Journal* 3, no. 2 (June 2014): 159–164.

Sinkey, Michael, and Trevon Logan. "Does the Hot Hand Drive the Market? Evidence from College Football Betting Markets." *Eastern Economic Journal* 40, no. 4 (Sept. 2014): 583–603.

Smith, Gary. "Horseshoe Pitchers' Hot Hands." *Psychonomic Bulletin & Review* 10, no. 3 (Sept. 2003): 753–58.

Smith, Gary, Michael Levere, and Robert Kurtzman. "Poker Player Behavior After Big Wins and Big Losses." *Management Science* 55, no. 9 (Sept. 2009): 1547–55.

Stern, Hal S., and Carl N. Morris. "A Statistical Analysis of Hitting Streaks in Baseball: Comment." *Journal of the American Statistical Association* 88, no. 424 (Dec. 1993): 1189–94.

Stöckl, Thomas, et al. "Hot Hand and Gambler's Fallacy in Teams: Evidence from Investment Experiments." *Journal of Economic Behavior & Organization* 117 (Sept. 2015): 327–39.

Stone, Daniel F. "Measurement Error and the Hot Hand." *American Statistician* 66, no. 1 (2012): 61–66.

Stone, Daniel F., and Jeremy Arkes. "March Madness? Underreaction to Hot and Cold Hands in NCAA Basketball." *Economic Inquiry* 56, no. 3 (July 2018): 1724–47.

Suetens, Sigrid, Claus B. Galbo-Jørgensen, and Jean-Robert Tyran. "Predicting Lotto Numbers: A Natural Experiment on the Gambler's Fallacy and the Hot-Hand Fallacy." *Journal of the European Economic Association* 14, no. 3 (June 2015): 584–607.

Sun, Yanlong, and Hongbin Wang. "Gambler's Fallacy, Hot Hand Belief, and the Time of Patterns." *Judgment and Decision Making* 5, no. 2 (Apr. 2010): 124–32.

———. "The 'Hot Hand' Revisited: A Nonstationarity Argument." *PsyCh Journal* 1, no. 1 (June 2012): 28–39.

Sun, Yanlong, and Ryan D. Tweney. "Detecting the 'Hot Hand': A Time Series Analysis of Basketball." Paper presented at the Forty-First Annual Meeting of the Psychonomic Society, New Orleans, LA, 2000.

Sundali, James, and Rachel Croson. "Biases in Casino Betting: The Hot Hand and the Gambler's Fallacy." *Judgment and Decision Making* 1, no. 1 (2006): 1–12.

Tversky, Amos, and Daniel Kahneman. "Belief in the Law of Small Numbers." *Psychological Bulletin* 76, no. 2 (1971): 105–10.

———. "Judgment Under Uncertainty: Heuristics and Biases." *Science* 185, no. 4157 (Sept. 27, 1974): 1124–31.

Tversky, Amos, and Thomas Gilovich. "The Cold Facts About the 'Hot Hand' in Basketball." *Chance* 2, no. 1 (1989): 16–21.

———. "The 'Hot Hand': Statistical Reality or Cognitive Illusion?" *Chance* 2, no. 4 (1989): 31–34.

Van Tilborgh, Louis, et al. "Weave Matching and Dating of Van Gogh's Paintings: An Interdisciplinary Approach." *Burlington Magazine* 154, no. 1307 (Feb. 2012): 112–22.

Van Tilborgh, Louis, Teio Meedendorp, and Oda van Maanen. "'Sunset at Montmajour': A Newly Discovered Painting by Vincent van Gogh." *Burlington Magazine* 155, no. 1327 (Oct. 2013): 696–705.

Vergin, Roger C. "Overreaction in the NFL Point Spread Market." *Applied Financial Economics* 11, no. 5 (2001): 497–509.

Wardrop, Robert L. "Simpson's Paradox and the Hot Hand in Basketball." *American Statistician* 49, no. 1 (Feb. 1995): 24–28.

———. "Statistical Tests for the Hot-Hand in Basketball in a Controlled Setting." Working paper, 1999. Available at pages.stat.wisc.edu/~wardrop/papers/tr1007.pdf.

Wilke, Andreas, and H. Clark Barrett. "The Hot Hand Phenomenon as a Cognitive Adaptation to Clumped Resources." *Evolution and Human Behavior* 30, no. 3 (May 2009): 161–69.

Wilke, A., and R. Mata. "Cognitive Bias." In *Encyclopedia of Human Behavior,* edited by V. S. Ramachandran. Amsterdam: Elsevier, 2012, 531–35.

Xu, Juemin, and Nigel Harvey. "Carry on Winning: The Gamblers' Fallacy Creates Hot Hand Effects in Online Gambling." *Cognition* 131, no. 2 (May 2014): 173–80.

Yaari, Gur, and Gil David. "'Hot Hand' on Strike: Bowling Data Indicates Correlation to Recent Past Results, Not Causality." *PLoS ONE* 7, no. 1 (2012): doi.org/10.1371/journal.pone.0030112.

Yaari, Gur, and Shmuel Eisenmann. "The Hot (Invisible?) Hand: Can Time Sequence Patterns of Success/Failure in Sports Be Modeled as Repeated Random Independent Trials?" *PLoS ONE* 6, no. 10 (2011): doi.org/10.1371/journal.pone.0024532.

Yuan, Jia, Guang-Zhen Sun, and Ricardo Siu. "The Lure of Illusory Luck: How Much Are People Willing to Pay for Random Shocks." *Journal of Economic Behavior & Organization* 106 (Oct. 2014): 269–80.

ARTICLES

Abbott, Henry. "Hot and Heavy: About NBA Shooting." *TrueHoop* (blog), April 17, 2009, espn.com/blog/truehoop/post/_/id/6241/hot-and-heavy-about-nba-shooting.

Abnos, Alex, and Dan Greene. "Boomshakalaka: The Oral History of *NBA Jam.*" *Sports Illustrated,* July 6, 2017.

Al-Kuttab, Yasmine. "Once a Sculptor for Saddam Hussein's Regime, Now Living in Abu Dhabi." *National* (United Arab Emirates), March 14, 2013.

Anger, Per. Oral history interview conducted by Joan Ringelheim. United States Holocaust Memorial Museum Collection, January 19, 1995, collections.ushmm.org/search/catalog/irn504796.

Appelbaum, Binyamin. "Streaks Like Daniel Murphy's Aren't Necessarily Random." *New York Times,* October 27, 2015.

Ayton, Peter. "Fallacy Football." *New Scientist,* September 19, 1998.

Barron, James. "He's Auctioned the 1776 Declaration, Twice." *New York Times,* July 4, 2000.

Base, Ron. "Fathers of the Princess Bride." *Toronto Star,* September 26, 1987.

"Bernt Hagen." In *Compendium of History and Biography of Polk County, Minnesota,* edited by R. I. Holcombe and William H. Bingham. Minneapolis, MN: W. H. Bingham, 1916, 312–13.

Bomsdorf, Clemens. "'Fake' Van Gogh in Attic Turns Out to Be Real." *Wall Street Journal,* September 10, 2013.

Borman, Stu. "A Chemistry Spy Story." *Chemical & Engineering News,* February 18, 2013.

Bradshaw, Lauren. "Career-Spanning Interview with Director Rob Reiner." *Cloture Club* (blog), July 22, 2014, clotureclub.com/2014/07/career-spanning-interview-director-rob-reiner-princess-bride-stand-etc.

Callahan, Tom. "When You're Hot, You're Hot . . ." *Time,* June 1988.

Cohen, Ben. "The Basketball Team That Never Takes a Bad Shot." *Wall Street Journal,* January 30, 2017.

———. "Does the 'Hot Hand' Exist in Basketball?" *Wall Street Journal,* February 27, 2014.

———. "The Golden State Warriors Have Revolutionized Basketball." *Wall Street Journal,* April 6, 2016.

———. "The 'Hot Hand' Debate Gets Flipped on Its Head." *Wall Street Journal,* September 28, 2015.

———. "Moneyball 2.0: Students in Harvard Club Prep to Be GMs." ThePostGame, February 24, 2011, thepostgame.com/features/201102/moneyball-20-students-harvard -club-prep-be-sports-gms.

———. "Stephen Curry's Science of Sweet Shooting." *Wall Street Journal,* December 17, 2014.

Dahdouh, Mary. "He Wed Science with Art, Solving Mystery of 'Sunset at Montmajour.'" *Houston Chronicle,* September 23, 2013.

Davidson, Adam. "Boom, Bust or What?" *New York Times Magazine,* May 2, 2013.

Davison, Drew. "Justin Grimm Proves Up to Challenge for Texas Rangers." *Fort Worth Star-Telegram,* June 17, 2012.

Dembart, Lee. "Logic Says Lady Luck Is Mere Illusion, but Statistics Fail to Entice." *Los Angeles Times,* July 25, 1988.

Drury, Paul. "The Making of NBA Jam." *Retro Gamer,* May 2013.

Durrett, Richard. "Justin Grimm Gives Quality Father's Day Gift." *Texas Rangers Report* (blog), ESPN, June 16, 2012, espn.com/blog/dallas/texas-rangers/post/_/id /4885728/justin-grimm-gives-quality-fathers-day-gift.

Ellenberg, Jordan. "'Hot Hands' in Basketball Are Real." *Slate,* October 26, 2015.

———. "The Psychology of Statistics." *Slate,* November 2, 2015.

"The Most Unlikely College Basketball Result of 2010." *Harvard Sports Analysis Collective* (blog), August 6, 2010, harvardsportsanalysis.org/2010/08/the-most-unlikely -college-basketball-result-of-2010.

Fields, Scott. "The Boy of Summer." *UCLA Magazine,* October 1, 2010.

Fleming, David. "Stephen Curry: The Full Circle." *ESPN the Magazine,* April 23, 2015.

Forsberg, Myra. "Rob Reiner Applies the Human Touch." *New York Times,* October 18, 1987.

Foster, Clint. "Justin Grimm Achieves Childhood Dream." *Texas Rangers Report* (blog), ESPN, June 16, 2012, espn.com/blog/dallas/texas-rangers/post/_/id/4885490 /justin-grimm-achieves-childhood-dream.

Freeman, Rick. "Kaplan's Project Might Change Face of Baseball." *Trenton Times,* January 1991.

Furlong, Ray. "Wallenberg Family Mark Centenary with Plea for Truth." BBC News, August 4, 2012.

Gatehouse, Gabriel. "Baghdad Diary: Saddam's Sculptor Makes Comeback." BBC News, June 16, 2010.

Gelman, Andrew. "Hey—Guess What? There Really Is a Hot Hand!" *Statistical Modeling,*

Causal Inference, and Social Science (blog), July 9, 2015, statmodeling.stat.columbia
.edu/2015/07/09/hey-guess-what-there-really-is-a-hot-hand.

Gibson, Lydialyle. "Return on Principles." *University of Chicago Magazine,* January–
February 2009.

Gleick, James. "'Hot Hands' Phenomenon: A Myth?" *New York Times,* April 19, 1988.

Goldsberry, Kirk. "DataBall." *Grantland,* February 6, 2014, grantland.com/features
/expected-value-possession-nba-analytics.

Goins, Richard. "Now Pitching New Statistics: Ari Kaplan." *Baseball America,* March 21,
1993.

Goldsberry, Kirk. "DataBall." *Grantland,* February 6, 2014, https://grantland.com
/features/expected-value-possession-nba-analytics.

Guth, Robert A. "Chicago Business School Gets Huge Gift." *Wall Street Journal,* November 7, 2008.

Haberstroh, Tom. "He's Heating Up, He's on Fire! Klay Thompson and the Truth
About the Hot Hand." ESPN, June 11, 2017, espn.com/nba/story/_/page/presents
-19573519/heating-fire-klay-thompson-truth-hot-hand-nba.

Harris, Scott. "Caltech Student Has the Stats to Make It to the Major Leagues." *Los Angeles Times,* June 6, 1990.

Hayes, Tim. "Grimm in the Bigs: The Timeline." *Bristol Herald Courier,* June 16, 2012.

"How Do You Spot a Real Van Gogh?" *Economist,* September 24, 2013.

Jenkins, Lee. "Stephen Curry's Next Stage: MVP Has Warriors Closing in on the NBA
Finals." *Sports Illustrated,* May 20, 2015.

Jensen, Jon. "Meet the Man Who Sculpted Saddam Hussein." CNN, May 3, 2013, cnn
.com/2013/05/01/world/meast/iraqi-sculptor-saddam-hussein/index.html.

Johnson, Albin E. "What I Saw in Sweden." *Rotarian,* September 1944.

Johnson, George. "Gamblers, Scientists and the Mysterious Hot Hand." *New York Times,*
October 17, 2015.

Kabil, Ahmed. "How Warren Buffett Won His Multi-Million Dollar Long Bet." *Medium,*
February 17, 2018, medium.com/the-long-now-foundation/how-warren-buffett-won
-his-multi-million-dollar-long-bet-3af05cf4a42d.

Keller, Bill. "Soviets Open Prisons and Records to Inquiry on Wallenberg's Fate." *New
York Times,* August 28, 1990.

King, Susan. "'The Princess Bride' Turns 30: Rob Reiner, Robin Wright, Billy Crystal
Dish About Making the Cult Classic." *Variety,* September 25, 2017.

Kuttler, Hillel. "Sabermetrician Ari Kaplan Uses the Science of Balls and Strikes to Illuminate the Fate of Holocaust Rescuer Raoul Wallenberg." *Tablet,* October 2, 2017,
tabletmag.com/jewish-news-and-politics/245911/sabermetrics-ari-kaplan-raoul
-wallenberg.

Lee, Ashley. "NYFF 2012: Rob Reiner, Billy Crystal, Robin Wright Spill Secrets About
the Making of 'Princess Bride.'" *Hollywood Reporter,* October 3, 2012.

Lee, Dave. "How Random Is Random on Your Music Player?" BBC News, February 19,
2015.

Lester, Elenore, and Frederick E. Werbell. "The Lost Hero of the Holocaust: The Search for Sweden's Raoul Wallenberg." *New York Times Magazine,* March 30, 1980.

Levy, Steven. "Requiem for a Shuffle." *Wired,* August 2, 2017.

Lewine, Edward. "There's a Method to My Desk's Madness." *New York Times,* May 18, 2013.

Lewis, Michael. "The King of Human Error." *Vanity Fair,* December 2011.

Looker, Dan. "Up Beet." *Successful Farming,* March 2003.

Loomis, Carol J. "Buffett's Big Bet." *Fortune,* November 2009.

Lopez, Michael, and Sadie Lewis. "An Exploration of MLB Umpires' Strike Zones." *Hardball Times,* May 4, 2018, tht.fangraphs.com/an-exploration-of-mlb-umpires-strike-zones.

Lowe, Zach. "Biting the Hot Hand: Basketball's Enduring Streakiness Debate Rages On." *Grantland,* September 20, 2013, grantland.com/the-triangle/biting-the-hot-hand-basketballs-enduring-streakiness-debate-rages-on.

Lowenstein, Roger. "Why Buffett's Million-Dollar Bet Against Hedge Funds Was a Slam Dunk." *Fortune,* May 11, 2016.

Madden, Michael. "Chilling the Hot Hand." *Boston Globe,* September 10, 1985.

Mahany, Barbara. "Captive Truth: U. of C. Chemist Works to Liberate the Facts on Raoul Wallenberg." *Chicago Tribune,* October 1, 1991.

McKean, Kevin. "The Orderly Pursuit of Pure Disorder." *Discover,* January 1987.

———. "When You're Hot, You're Not." *Discover,* June 1985.

McWeeny, Drew. "The M/C Interview: Rob Reiner Talks 'Flipped,' 'Princess Bride,' 'Misery' and More." *HitFix,* August 4, 2010.

Michaud, Christopher. "Magna Carta Fetches $21.3 Million at Sotheby's Auction." *Reuters,* December 18, 2007.

Miller, Joshua, and Adam Sanjurjo. "Momentum Isn't Magic: Vindicating the Hot Hand with the Mathematics of Streaks." *Conversation,* March 28, 2018.

Morrissey, Kate. "Iraqi Sculptor Hopes for New Life in U.S." *San Diego Union-Tribune,* November 23, 2017.

Preston, Julia. "Big Disparities in Judging of Asylum Cases." *New York Times,* May 31, 2007.

Reifman, Alan. "Transcript from Tom Gilovich Online Chat." *Hot Hand in Sports,* September 2002.

Remnick, David. "Bob Dylan and the 'Hot Hand.'" *New Yorker,* November 9, 2015.

Rottenberg, Josh. "*The Princess Bride:* An Oral History." *Entertainment Weekly,* October 14, 2011.

Segal, Julie. "David Rubenstein's Monopoly Money." *Institutional Investor,* May 4, 2017.

Seides, Ted. "Why I Lost My Bet with Warren Buffett." *Bloomberg,* May 3, 2017.

Shen, Allan. "Renowned Mathematician and Professor Elias Stein Passes Away at 87." *Daily Princetonian,* February 5, 2019.

Silverman, Jeff. "Roundabout Director Rob Reiner Makes His Movies Work—Somehow." *Chicago Tribune,* October 4, 1987.

Singal, Jesse. "How Researchers Discovered the Basketball 'Hot Hand.'" *New York*, August 14, 2016.

Smith, Hedrick. "U.S. Exchanges 2 Russian Spies for 2 Americans." *New York Times*, October 12, 1963.

Sommer, Jeff. "Challenging Management (but Not the Market)." *New York Times*, March 16, 2013.

Stromberg, Joseph. "Scientists Dismissed 'Hot Streaks' in Sports for Decades. They Were Wrong." *Vox*, June 3, 2015, vox.com/2015/6/3/8719731/hot-hand-fallacy.

Sullivan, Jeff. "Incredulous Responses to Bill Miller's Strike Zone." FanGraphs, October 30, 2017, blogs.fangraphs.com/incredulous-responses-to-bill-millers-strike-zone.

Tully, Shawn. "How the Really Smart Money Invests." *Fortune*, July 6, 1998.

Vinocur, John. "Swedish Hero Is in Soviet, Panel Says." *New York Times*, January 16, 1981.

Voss, Greg. "Sneaking Up on Success: An Interview with Mark Turmell." *Softline*, November 1981.

Wallace, Anise C. "Perils and Profits of Pension Advisers." *New York Times*, September 11, 1983.

Warren, Jamin. "Mark Turmell." *Kill Screen*, May 12, 2011.

Wolff, Alexander. "The Olden Rules." *Sports Illustrated*, November 25, 2002.

Yeh, Molly. "In Season: Sugar Beets." *Modern Farmer*, October 23, 2013.

Zimmer, Carl. "That's So Random: Why We Persist in Seeing Streaks." *New York Times*, June 26, 2014.

Zweig, Jason. "Making Billions with One Belief: The Markets Can't Be Beat." *Wall Street Journal*, October 20, 2016.

SHOWS

"Bill Miller (Major League Baseball Umpire)." *Off the Lip Radio Show*, January 24, 2017, youtu.be/eaWM1xoNw14.

"David Booth Interview." Index Fund Advisors, October 12, 2016, youtu.be/LWC GYz44hac.

"Dean's Distinguished Speaker Series: David Booth." Hosted by Judy Olian. UCLA Anderson School of Management, January 29, 2014, youtu.be/pZcXeI2ke6o.

"An Interview with the Chicago Team Who Is Investigating the Fate of Raoul Wallenberg." By Lisa Pevtzow. Jewish United Fund, April 8, 2013, juf.org/interactive/vPlayer.aspx?id=420403.

Kestenbaum, David, and Jacob Goldstein. "Brilliant vs. Boring." Episode 688, *Planet Money* (podcast), NPR, January 23, 2019, npr.org/sections/money/2019/01/23/688018907/episode-688-brilliant-vs-boring.

Meedendorp, Teio. "Discovering Vincent van Gogh's *Sunset at Montmajour*." TEDx Talks, University of St. Andrews, June 16, 2014, youtu.be/SyzdA_dQjD0.

"Morning Session—2006 Meeting." Warren Buffett Archive, CNBC, May 6, 2006, buffett.cnbc.com/video/2006/05/06/morning-session---2006-berkshire-hathaway-annual-meeting.html.

Ritholtz, Barry. "Thomas D. Gilovich Talks About Human Behavior." *Masters in Business* (podcast), *Bloomberg,* January 25, 2018, soundcloud.com/bloombergview/thomas -d-gilovich-talks-about-human-behavior.

"Special Episode: Sugar Beet Harvest." *America's Heartland,* January 26, 2011, youtu.be /ksN7h-ZpFWc.

"VIP Distinguished Speaker Series: David Booth, CEO of Dimensional Fund Advisors." McCombs School of Business, University of Texas, Austin, February 26, 2013, youtu .be/cCp1m7rG0Q0.

ACKNOWLEDGMENTS

There are so many things that writing a book for the first time has taught me. One of them is that it's almost certainly impossible to thank everyone who makes a first book possible. Here goes anyway.

My wonderful agent, Eric Lupfer, turned an idea into a proposal, a proposal into a book, a book into a much better book. Eric often knew what I was trying to write before I did, and then he gently made sure I wrote it. I'm deeply grateful to have him in my corner. The same goes for Christy Fletcher and everyone at Fletcher & Company.

I'm incredibly fortunate that my first book was in the care of Geoff Shandler. Geoff had a vision for what this book should be and pushed me to bring that vision to life even when I wasn't entirely sure that I could. If you're looking for his deft touch, you can find it on every page. It was such a privilege to be edited by him. Once he whipped my copy into shape, Geoff's team at Custom House and HarperCollins got to work. Kayleigh George, Maureen Cole, Jessica Rozler, Nancy Tan, Leah Carlson-Stanisic, Ploy Siripant, Trina Hunn, Liate Stehlik, and Ben Steinberg turned a Google Doc into the book you're reading. The fantastic Molly Gendell made sure of it. Hank Tucker plowed through hundreds of academic papers and translated them into English all while he was still in college. John Vilanova was my fact-checker extraordinaire. He picked apart the manuscript and did a truly remarkable job making sure this book was actually true. Any remaining errors are mine alone.

The *Wall Street Journal* is where I learned how to be a reporter, and there was no one more formative in this education than the brilliant Sam Walker. This is a frightening prospect to anyone who knows Sam, but it has made all the difference for me. I have no idea what I would be doing if not for Sam, but I do know that it wouldn't be nearly as fun. I'm so lucky that he gave me the break of a lifetime. I'm even luckier to have Bruce Orwall as my *Journal* rabbi now. Bruce is a total dream of an editor. He is wise. He is patient. He not only allows me to pursue crazy ideas but actively encourages them—and then he makes them seem only slightly less crazy. I'm confident this book wouldn't exist without his unwavering support. I really hope that one of his sports teams gets hot at some point in the next century.

I'm also indebted to Mike Miller, Matt Murray, and many fantastic colleagues across the *Journal*'s mighty and inspiring newsroom, most of all the people directly responsible for the paper's sports coverage at this writing: Rachel Bachman, Brian Costa, Jim Chairusmi, Jared Diamond, Tongan sports correspondent Joshua Robinson, and the indefatigable Andrew Beaton. Did I forget anyone? Oh, right. The completely unforgettable Jason Gay. Jason is menschy, generous, and right about everything always. He also happens to be the funniest writer on the planet. When I tell people I write about sports for the *Journal*, it's almost inevitable they will ask if I'm Jason Gay. I wish. He's the greatest mentor and role model I can imagine.

There were a few people who went above and beyond to read early drafts of this book. Jonathan Clegg has the highly useful habit of improving everything I write once I beg him to read it, and he loaned his wordsmithing expertise to the only thing I've ever filed that was shorter than he expected. I was reminded why I was so desperate for Geoff Foster's eye when he pinpointed the exact parts of this book that needed the classic Foster treatment. Sam Walker's edit was both terrifying and thrilling—the full Sam Walker experience. In addition to being a phenomenal friend, John Harpham is the smartest person I know. I'm excited to read all the books that he's going to write. Andrew Yaffe has read more of my writing than anybody, at least if you count Gchats, and his reward for putting up with me for so long was being forced to read even more of it.

Thank you to everyone who asked how the book was coming along, and thank you to everyone who knew not to ask. Among the many friends who provided all the support and distractions I needed to actually finish: Scott Cacciola, Kevin Clark, Marc Tracy, Joe Coscarelli, Nate Freeman, Dan Romero, Sam Schlinkert, Katie Baker, J.R. Moehringer, Alana Newhouse, Bari Weiss, the *Unorthodox* crew, the photographer Samantha Bloom, Andrew and Christen Yaffe, Greg and Olivia Beaton, Matthew Futterman, and Amy Einhorn, Matthew Henick and Alaina Killoch, and so many other friends who I'm sure will be nice enough not to remind me that I've missed them.

Family! You all know who you are. Shout-out to the entire Butnick-Rothaus clan. Cliff, Francesca, and Noah Silverman, and Howard and Elyse Butnick will always deserve special mentions. The Resnikoffs have been by my side since books terrified me. Sara and David Silver keep me sane and support me even when I spend too much time staring at my phone. Wendy and Jesse Cohen have always been my biggest fans and most loyal readers. They are the people who made everything possible—including my unfortunate basketball career. Thank you.

Stephanie Butnick is the reason I feel like I have the hot hand every day. She lived with this book from the very beginning to the very end, editing it throughout, pushing me and championing me and protecting me from an occasionally psychotic cat the whole time. The first sentence in this book was for you. Now the last one is, too.

INDEX

A+ CinemaScore, 53
active management, 116
Allen, Kyle, 4–6
Al-Saffar, Alaa
    application as pending, 152
    as asylum applicant, 128, 143, 147–153
    background, 126
    death threat, 127
    demonstration of fear of death and, 150
    Hussein commission, 126–127
    interview with asylum officer, 150
    issues with work in new life, 149
    in Los Angeles, 125, 128
    memorial sculpture commission, 127
    relocation, 148
    Saddam Hussein and, 125–127, 150
    uncertainty, 151–152
    visit with, 148–149
    working from home, 148
Al-Saffar, Zinah, 126–128, 152
Antony and Cleopatra, 41, 63
Apple, Smart Shuffle, 68
Apple II, 13, 14
arrogance, 56
artists and scholars
    Albert Einstein, 49, 54
    hot hand in, 41–64
    peaking of, 55
    Rebecca Clarke, 42–49
    Rob Reiner, 49–54
    William Shakespeare, 41–42

asylum applicants
    Alaa Al-Saffar, 128, 143, 147–153
    cruelty of process, 144
    heard twice, 147
    judges and, 144–148
    "Refugee Roulette" paper and, 143–144
    wheel of fortune and, 144
asylum judges
    Bruce Einhorn, 145
    decision analysis, 146
    gambler's fallacy and, 146–147
    in refugee roulette, 143–148
    reviewing decisions, 147
asylum system, 151–152
Atkinson, Ron (Big Ron), 141–142
Auerbach, Red, 92, 244
Ayton, Peter, 139–141, 143

bad shots, 5–6
Barea, J. J., 168
Barrett, H. Clark
    collaboration with Wilke, 81–82
    in Ecuador, 82
    UCLA students and Shuar foragers and,
        82–83
basketball
    founding of, 97–98
    "gravity," 34
    naming of sport, 98
    NBA algorithm change, 24
    pick-up, 244

basketball (*cont.*)
  player behavior change and, 203
  randomness in, 79
  referees, 165–166
  rest of the world and, 8
  rules of, 119–123
  SportVU and, 164–170
  success formula, 26
  three-point shot and, 24–25
  Tom Gilovich and, 78–80
  Tversky, Amos and, 78–80
basketball hot hand. *See also* hot hand
  author high school story, 1–2
  as Bigfoot, 8
  Gilovich, Vallone, and Tversky paper
    and, 89–92
  Gilovich and Tversky and, 79–80
  NBA fans and, 87–88
  Pine City Dragons, 4–7
  player behavior change and, 200–201
  player likelihood of shooting with, 203–204
  Stephen Curry and, 23–37, 51
  three-point shoot-out and, 216–218
Bonger, Andries, 220, 233
Bonger List, 220, 221, 230, 233
Booth, David
  background, 98
  betting against the hot hand and, 102
  comparative advantage and, 102
  Dimensional Fund Advisors and, 113,
    114, 117
  dissertation, 98–99
  diversification and, 114
  Eugene Fama and, 99–101, 112, 114, 245
  first office, 111
  market efficiency and, 100–101
  as Ph.D. student, 99
  principles over patterns and, 113
  purchase of original rules of basketball, 123
  Rex Sinquefield and, 111–113
  on streaks as aberrations, 113
  University of Chicago and, 99–102, 118

  as wealth celebrity, 117–118
  wealth on single idea, 117
  working assumption of "no" and, 118
Boston Celtics, 79, 88–89, 92, 168, 244
Bubble Safari, 38–39
Buffett, Warren, 108, 115, 116–117

Carlberg, Ingrid, 159
Carril, Pete, 26
casinos
  asylum courts as, 144
  decision-making and, 136
  roulette in, 136–143
Chen, Daniel, 145
"circumstance," 48
Clarke, Rebecca
  Anthony Trent and, 43, 44, 45
  background, 42–43
  "circumstance" and, 48
  before death, 48
  at eighty-nine years old, 45
  Elizabeth Coolidge and, 44–45, 46
  fresh recordings, 48
  on hot hand, 46
  on intensity and concentration, 47
  interviews with and essays about, 48
  memoir, 46
  Morpheus performance, 43
  placing second and, 44–45
  relationship with her past, 46
  Robert Sherman and, 45–48
  sonata, reviews of, 48
  sonata, writing, 44
  sonata performance at Lincoln Center, 48
  tarot reading, 48–49
clumps, 81, 83, 85–86
cognitive bias, 88
cognitive illusion, 91, 115, 213
cognitive pattern, hot hand as, 81
Cognitive Psychology, 92
coin tosses, 82–83, 223, 224–225
comparative advantage, 102

Complex Heat, 200, 203
confidence, 30, 56
Coolidge, Elizabeth, 44–45, 46
Cornell basketball teams, 88–89
Croson, Rachel
     background, 136
     fifty-fifty bets in roulette and, 138, 142–143
     gambler's fallacy and, 141–142
     Ph.D. student security tapes and, 138
     "seem to believe" and, 143
Cuban, Mark, 167–168, 191
Curry, Dell, 17, 18
Curry, Stephen
     as always the smallest player, 17
     as best NBA shooter, 16–17
     breakthroughs, 37
     defense of, 33–34
     drafting of, 25
     fans watching him get hot, 36–37
     February 27, 2013 and, 16, 17, 19, 23–37
     on fire (2/27/13), 30
     first three-point shots (2/27/13), 29–30
     first time in Garden, 18
     as foundational piece of roster, 28
     hot hand, 23–37, 51
     last shot (2/27/13), 35
     management encouragement to shoot, 36
     NBA Jam and, 23, 39, 247
     ovation from opposing fans, 35
     Pacer fight and fine, 29
     Queensway Christian College and, 32
     relearning how to shoot, 18
     second half (2/27/13), 34–35
     slingshot as weapon, 18–19
     sound of shots (2/27/13), 35
     taking low-percentage shot (2/27/13), 29–30
     on third bus, 23, 24
     three-point shoot-out and, 216–217

Dallas Mavericks, 167–168
Dennis' Place for Games, 20, 22, 39
Dimensional Fund Advisors, 113, 114, 117

Eichmann, Adolf, 160–161
Einhorn, Bruce, 145
Einstein, Albert, 49, 54
Ellenberg, Jordan, 226
Elway, John, 78
Emancipation Proclamation, 119, 120, 122
embracing the hot hand, 51
exam grading, 147
Ezekowitz, John
     Brian Kopp and, 198
     Carolyn Stein and, 192–193
     Gilovich, Vallone, and Tversky paper and, 192
     in HSAC, 187–188, 191
     Mark Cuban and, 191
     Phoenix Suns and, 191–192
Ezekowitz and Stein study
     Complex Heat, 200, 203
     findings, skepticism of, 204
     Gilovich and Tversky and, 204–205
     hot hand data, 202
     hot-hand fallacy, 204
     hot hand topic and, 198
     Larry Summers and, 204
     player behavior change and, 202–203
     shot difficulty and, 203
     Simple Heat, 200
     SportVU data and, 199
     starting of, 196

faking shots, 2
fallacy. See gambler's fallacy; hot-hand fallacy
Fama, Eugene, 99–101, 112, 114
Fischer, Ilan, 139–140, 143
Fisher-Yates algorithm, 66
"flow states," 31

gambler's fallacy
     asylum judges and, 145–147
     baseball umpires and, 145
     believing in both the hot-hand fallacy
          and, 139–140, 142, 143
     defined, 138

gambler's fallacy (*cont.*)
  hot-hand fallacy versus, 139
  winning and losing and, 142–143
Gelman, Andrew, 214–215, 235–238
Gilovich, Tom
  basketball and, 78–80
  as child, 70
  experiment with Matt and Max (students), 245–249
  Ezekowitz and Stein study and, 205
  meeting at Cornell, 243–244
  Philadelphia 76ers and, 86–88
  on pick-up basketball, 244
  as roadie to academic stars, 71
  Stanford and, 71–73
  Tversky and, 77–80
Gilovich, Vallone, and Tversky paper
  alternation rate, 90–91
  data quality, 199–200
  explainer of paper, 190
  Ezekowitz as having read, 192
  kids at Harvard and, 190
  mathematical flaw, 226
  replication in soccer, 140
  series of runs and responses, 89–90
  shooters in, 212
  shot selection, 197
  writing of, 91–92
Golden State Warriors, 23–24, 25, 27–29, 39, 168
Goldman, William, 52
Goldsberry, Kirk, 199
good shots
  as close to the basket, 25
  Pine City Dragons and, 5–6
"gravity," 34
Grimm, Justin
  advantage, 132–133
  called up to Texas Rangers, 129
  in Double-A ball, 128–129
  filthy pitch, 134
  first major league pitch, 129

goal to strike out, 131
home run off of, 135
Jed Lowrie and, 129–132, 134–135
Gulick, Luther, 96–97

Hagen, Bernt, 108, 109
Hagen, Nick
  background, 102–103
  beet harvest preparation, 107
  as farmer disguised as a musician, 103–104
  on farming as defense, 108
  farming education, 105
  fear of the weather, 108–109
  going back to the farm, 104–105
  hot hand and, 109
  lack of control, 110–111
  margin for error, 110
  mitigating risk, 107
  Molly Yeh and, 104–105
  plan B and, 105–106
  on playing field constantly shifting, 109
  playing the long game and, 105
  preparation for the worst, 106, 111
  prices going up and, 105
  on principles over patterns, 110
  on success, 110
  as trombone player, 106
  on wisdom of Warren Buffett, 108
Harvard Sports Analysis Collective (HSAC)
  Carolyn Stein and, 187
  existence of, 187
  Gilovich, Vallone, and Tversky paper and, 190
  John Ezekowitz and, 187–191
  kids in, 189–190
  Moneyball and, 188–189
Hayden, Ben
  background, 83–84
  data and, 84
  monkey study, 84–86
"heat checks," 51
"Hey—Guess What? There Really Is a Hot Hand," 236

Hodges, Craig, 217, 223–224, 226–227
hot hand. *See also* basketball hot hand;
    specific hot hand studies and papers
    acknowledging existence of, 247
    artists and scholars, 41–64
    author high school story, 1–2, 3
    belief in, 9, 91–92, 101
    bias, 80
    as cognitive pattern, 81
    Dashun Wang research, 54–56
    defined, 2
    embracing, 51
    ethereal feeling of, 2
    feeling of, 2–3
    happiness and, 31
    in human thinking, 83
    in monkey study, 84–86
    studies about, 3–4, 8
    success begets success and, 51
    truth about, 246
    viewing everywhere, 8
hot-hand fallacy
    believing in both gambler's fallacy and,
        139–140, 142, 143
    Ezekowitz and Stein study, 204
    gambler's fallacy versus, 139
    Miller and Sanjurjo and, 214
    variation of changes of making a shot
        and, 197
hot-hand period
    Dashun Wang on, 49, 55
    defined, 55
    knowing if you've already had, 55
Hussein, Saddam, 125–127, 150

investment fees, 116
"It's Okay to Believe in the 'Hot Hand,'" 190

Jacobs, Eli, 177
James, Bill, 177, 201–202
Jobs, Steve, 68
Johnson, Don, 229–231, 234

Kahneman, Daniel, 73–77, 238–239
Kahneman and Tversky
    American Psychological Association
        survey, 75
    first paper in *Psychology Bulletin*, 76
    law of small numbers, 77
    Nobel Prize and, 93
    on sabbatical at Stanford, 73
    Science paper, 76–77, 91
    Tom Gilovich and, 73, 77
Kaplan, Ari. *See also* Makinen and Kaplan
    background, 176
    Bill James and, 177
    Major League Baseball study, 176–177
    with Orioles, 177–178
    paper route, 179–180
King Lear, 41, 63
Kitzrow, Tim, 22, 30, 88
Knight, Phil, 122
Kopp, Brian
    background, 163
    Ezekowitz and, 198
    Lakers and, 186
    Mark Cuban and, 167–168
    mythical geeks and, 169–170
    after NBA lockout, 185
    NBA pitch, 165–166
    SportVU possibilities and, 169
    Tamir and Oz and, 164

Lackey, James, 32
Lacob, Joe, 28–29
Laplace, Pierre-Simon, 138–139
Larina, Varvara. *See also* Makinen and Kaplan
    first interview with, 178–181
    identification of Wallenberg, 180, 181
    odds of being wrong, 196
    remembrance of prisoner, 179–180
    second interview with, 181
    third interview with, 181
law of small numbers, 77
Lear, Norman, 53

Lowrie, Jed
  Astros trade for, 130–131
  background, 130
  goal to hit home runs, 131
  home run off of Grimm, 135
  Justin Grimm and, 129–132, 134–135
  Sig Mejdal and, 130

Macbeth, 41, 63
Madison Square Garden, 16, 18, 19, 23, 36, 109
Makinen, Marvin
  final conclusions on Wallenberg, 196
  frequency of prisoner moving and, 184–185
  hearing of Wallenberg's heroism, 171–173
  imprisonment recollection, 193
  interviews with the Swedish diplomats, 173
  Josif Terelya interview, 181–182
  as professor, 173
  reading "The Lost Hero of the Holo-
      caust," 173
  return to Vladimir Central Prison, 182–183
  Vandenberg and Wallenberg and, 174
  Varvara Larina and, 178–181
  in Vladimir prison, 172
Makinen and Kaplan
  algorithms, 185
  in avoiding human error, 184
  cell 25 and, 182, 194
  cell occupancy and, 195–196
  computer queries in Russia, 194–195
  the Fool and, 182
  Larina evidence, 178–181, 194–195
  life devotion to finding truth, 185
  objectivity, 184
  pattern of prisoner movement and, 196
  prisoner cell history and, 183
  prison map, 194
  Russian handwriting experts and, 184
  Terelya evidence, 181–182, 194
  travel to Vladimir Central Prison, 182–183
market efficiency, 100–101

Masur, Norbert, 157–158
Matt and Max (students), experiment with,
      245–249
McQuown, John "Mac," 112
Meedendorp, Teio, 222, 232–234
Mejdal, Sig, 130–131
Midway, 14–16, 19, 20–22
Miller, Bill. *See also* Grimm, Justin; Lowrie,
      Jed
  calls after two straight strikes, 135
  filthy pitch, 134
  as lifer behind the plate, 132
  PITCHf/x and, 133–134
  as umpire, 131–135
Miller, Josh
  Andrew Gelman and, 235
  background, 209–210
  Daniel Kahneman and, 238–239
  fire and, 210
  meeting Sanjurjo, 210
  watching Stephen Curry, 240–241
Miller and Sanjurjo
  in Autun, 228, 234
  background, 210–211
  in Betanzos, 212–213
  cognitive illusion and, 213
  coin flips, 223, 224–225
  controlled test of shooting, 211–212
  Craig Hodges and, 217, 223–224
  decision to work together, 211
  evidence of hot hand shooting, 215
  Gelman and, 214–215
  Gelman writing about, 236
  hot-hand fallacy and, 214
  mathematical bias and, 223
  results as inversely correlated and, 213
Miller and Sanjurjo paper
  Andrew Gelman and, 235–238
  as brilliant and infuriating, 226
  coin flips, 226
  Craig Hodges and, 226

findings of statistical bias, 226–227
  statistical bias detection, 227
  statistical formulas, 227
  tour, 236–238
Miller and Sanjurjo paper (published), 240
MIT Sloan Sports Analytics Conference, 186, 204
Moneyball (Lewis), 5, 188–189, 201, 245
monkey study, 84–86
Morpheus, 43
Morris, Carl, 188
Moskowitz, Toby, 145–147
Mustad, Christian
  background, 207
  buyer's remorse, 208
  death of, 221
  as deluded by own self doubt, 234
  as duped, 209
  painting in attic, 221
  Pellerin and, 208–209, 221
  Van Gogh purchase, 207–208
Myers, Bob, 27–28

Naismith, James
  background, 95–96
  failure, 97
  founding of basketball, 96–98
  ideal sport and, 97
  Luther Gulick and, 96–97
  rules of basketball, 119–123
NBA Jam. See also Turmell, Mark
  brain tendency to search for patterns and, 22
  Bubble Safari and, 38–39
  defined, 16
  machines sold, 21
  Mark Turmell and, 16, 19–23, 38–39
  player reward, 39
  quirk in game mechanics, 21
  Stephen Curry and, 23, 247
  success, 21–22
  test arcade, 20

usage statistics, 20
  voice-over role, 22
Nicastro, Neil, 20
non-randomness, 80

Osmak, Kirill, 179, 181, 195
Oz, Gal, 162–164

passive management, 116
Pellerin, Auguste, 208–209, 221
Philadelphia 76ers, 86–88
Phoenix Suns, 191–192
Pine City, Minnesota, 4–5
Pine City Dragons, 4–7
PITCHf/x, 133–134
plague
  background, 56–57
  quarantine and, 60
  Romeo and Juliet and, 58–61
  Shakespeare and, 57–64
  Shakespeare and competition and, 63–64
  Stratford-upon-Avon and, 57
  symptoms, 57–58
Poláček, Lukáš, 65, 68–69
Pollack, Harvey, 86–87
The Princess Bride, 51–54, 56
Princeton basketball, 26–27, 28
principles over patterns, 110, 111, 113

Queen Anne, 41, 63
Queensway Christian College, 32

randomness
  in basketball, 79
  learning about, 224
  seeing patterns in, 78
  Tversky and, 91
Rapoport, Ammon, 136, 137
Rebecca Clarke Reader, 48
Rebecca Clarke Society, 48
Redden, David, 118–123

Redford, Robert, 52
"Refugee Roulette," 143`
Reiner, Rob
  A+ CinemaScores, 53–54
  as critical and commercial success, 50–51
  A Few Good Men and, 53
  Normal Lear and, 53
  The Princess Bride and, 51–54, 56
  Stand by Me and, 50
  The Sure Thing and, 50
  This Is Spinal Tap and, 49–50
  When Harry Met Sally and, 53
  William Goldman and, 52
risk, mitigating, 106–107
Rivett, Jamie, 21
The Rocks, 230–231, 233, 234
Romeo and Juliet
  Friar John and, 59–60
  Friar Laurence and, 58–60
  plague and, 58–61
  Romeo's death, 58–59
  Romeo thinking Juliet is dead and, 59
Roosevelt, Eleanor, 170–171
Ross, Lee, 72–73, 77, 80
roulette tables, electronic scoreboards, 141–142
Rubenstein, David, 119–123
rules of basketball. See also Naismith, James
  auction of, 119–123
  bidders bidding on, 120–123

Sanjurjo, Adam. See also Miller and Sanjurjo
  meeting Miller, 210
  talk in Toulouse, 213
Schutzpass, 158–159
Science paper (Kahneman and Tversky), 76–77, 91
search for truth, 162
Seides, Ted, 115, 116–117
Shakespeare, William
  Antony and Cleopatra, 41, 63
  chronology of plays, 61

hot hand, 41–42, 61–64
  King Lear, 41, 63
  Macbeth, 41, 63
  plague and, 56–64
  Romeo and Juliet and, 58–61
  as a streaky writer, 62, 64
  success, 64
Shapiro, James, 61–62
Sherman, Robert, 45–48
shooting strategy, 7
Shuar foragers, 82–83
Shue, Kelly, 145–146
shuffling
  Fisher-Yates algorithm, 66
  Smart Shuffle, 68
  Spotify, 65–70
Simple Heat, 200
Sinquefield, Rex, 111–113
Smart Shuffle, 68
Sneakers (Turmell), 13, 14
SportVU
  ball height measurement, 166
  basketball and, 165–170
  Dallas Mavericks and, 167–168
  data, as messy, 198–199
  defined, 163
  first teams to purchase, 168
  NBA price tag, 168–169
  Oz and Tamir development of, 162–164
  professors as saviors of, 170
  as a promise, 166–167
  sales after NBA lockout, 185–186
  starting of, 163
  STATS purchase of, 164
  visual intelligence and, 163
Spotify
  Babar Zafar and, 69–70
  Fisher-Yates algorithm, 66
  Lukáš Poláček and, 68–69
  problem, 65, 67
  shuffling, 65–70
  workplace shuffle, 68

Stand by Me, 50

Stanford
Fred Terman and, 71
John Elway and, 78
Kahneman and Tversky on sabbatical at, 73
Lee Ross and, 72–73
Meet the Faculty seminar, 72–73
psychology department, 70–72, 99
Tom Gilovich and, 71–73

STATS (Sports Team Analysis and Tracking), 163–164, 167, 169

Stein, Carolyn, 186–187, 192–193. *See also* Ezekowitz and Stein study

success begets success, 51

Summers, Larry, 197, 204–205

Sundali, Jim
background, 136–137
bettors and prior numbers and, 141
fifty-fifty bets in roulette and, 138, 142–143
gambler's fallacy and, 142
Ph.D. student security tapes and, 138
"seem to believe" and, 143
study of gambling, 137

Sunset at Montmajour, 207, 209, 219–220, 233, 234

The Sure Thing, 50

"Surprised by the Hot Hand Fallacy? A Truth in the Law of Small Numbers" (Miller and Sanjurjo), 240

Taleb, Nassim Nicholas, 238

Tamir, Miky, 162–164

Terelya, Josif, 181–182, 194

Terman, Fred, 71

This Is Spinal Tap, 49–50

three-point shoot-out
Craig Hodges and, 217, 223
defined, 216
Miller and Sanjurjo and, 216–218
Stephen Curry and, 216–217

three-point shot
arc as boundary in time, 25

introduction of, 24
use statistics, 24–25

Trent, Anthony, 43, 44, 45

Turmell, Mark
Bubble Safari and, 38–39
first work with computers, 11–12
game ability and, 21
hot hand concept and, 23
meeting with, 37
Microsoft and, 14
at Midway, 14–16, 19, 22
NBA Jam and, 16, 19–23, 38–39
as pyromaniac, 11
Sneakers, 13
test audiences and, 15–16
Wozniak and, 13–14
at Zynga, 38

Tversky, Amos. *See also* Kahneman and Tversky
accolades, 74
background, 73–74
basketball and, 78–80
as beloved, 93
death of, 93
Gilovich, Vallone, and Tversky paper and, 89–92
Kahneman's eulogy for, 93
MacArthur Foundation Fellowship, 74
randomness and, 91
recognition of brilliance, 74
teaching at Stanford, 75, 78

UCLA basketball, 26, 27, 28

UCLA students, in Barrett study, 82–83

"Underestimating the Fog," 201

University of Chicago, 99–102, 118

Vallone, Robert, 87, 88, 89, 90–91

Van Gogh
in Arles, 218–220
assignment of numbers to work, 220
Autumn Landscape, 220–221